THE
ONENESS
OF
ECOLOGY

DONALD PRESTON

ALSO BY DONALD PRESTON

I Love You One Thousand Houses: A Memoir

ISBN-13: 978-1495237232

ISBN-10: 1495237230

Front cover: The Bubble Nebula (NGC 7635)
hubblesite.org
Photo selected by Victoria Baker

Book design by

PrescottSecretarial.com

In memory of my parents, Helen and Edward

ACKNOWLEDGMENTS

I am grateful to the scientists whose work makes the world a better place to live. They made this book possible.

Through the deep caves of thought I hear a voice that sings:—

Build thee more stately mansions, O my soul,
* As the swift seasons roll!*
* Leave thy low-vaulted past!*
Let each new temple, nobler than the last,
Shut thee from heaven with a dome more vast,
* Till thou at length are free,*
Leaving thine outgrown shell by life's unresting sea!

From "The Chambered Nautilus"
by Oliver Wendell Holmes, Sr.

CONTENTS

Preface

What brings me to the point of writing this book is a confluence of observations that caused fresh thinking for me.

First, I see that humans are overwhelmed by a vast amount of informational *input* (from TV, the Internet, and other technologies), yet we produce very little real, original *output*—especially *mental output.* The ratio of input to output is dreadfully out of balance.

Second, I see that our lives from birth are largely predetermined by race, gender, religion, custom, culture, social status, country of origin, and other rigid situations, but not by ourselves. These differences have caused disunity among humans from the very beginning.

To a great extent, both of these conditions or circumstances relegate us—our very lives—to be products of forces and influences outside of our personal control and choosing. *There is little or no self-determination.*

Who we are, from childhood through adulthood, is perhaps now more than ever shaped by "others." As consumers, we increasingly become more like the branded products we buy. Useful and valuable information is harder to find in the massive, intrusive yet often subtle marketing that demands our almost constant attention and consciousness.

As you read these pages, consider this metaphor: You are one of many little seeds planted into fertile soil. This soil, rich in nutrients, exists in an atmosphere of sun and moisture. There is a conscientious farmer or loving, Supreme Being caring for and looking after us.

The garden and lush soil is the planet Earth and the

farmer is our supreme being, God. He loves us all and wants us to grow and flourish. He mostly wants us to love Him by being good, productive human beings. Those that use this gift live life to its fullest.

We all know, of course, that in reality we are not "seeds." Many of us want to grow throughout our lives to please our Creator. To do this we must hold ourselves accountable for our actions and the thinking that drives our actions. While reading this book, you must bring (and use) logic, rationality, knowledge, and intelligence to the task. I ask that you not "filter" and judge what I say using what you already think you know.

In short, for you to hold yourself fully accountable, you must truly think for yourself, as an individual. You must allow for new ways of thinking. What is already in your mind will resist new information that won't fit into what is already familiar and comfortable for you.

A mental rebirth, like your actual first birth, is rarely easy and painless.

The ideas I present are about God, the universe, science and humanity, all within the realm of beliefs (new and old), facts, fiction, and theory. "I have come into the world as a light." (John 12:46) When I write about science, the Earth, the universe and humanity, I see a Supreme Being's hand in all of them.

At times I will tread on deeply held ideas and beliefs, and also introduce enlightening information. I humbly only ask that you try to think and feel comfortable on this voyage—this mental odyssey.

I hope the upcoming pages will begin a mental rebirth for you. Some of those around you may become suspicious of your changed outlook and fresh individualism. Some may ostracize you.

In a previous book, my memoir, I mentioned an old Scottish adage: "If you're walking in a crowd, you are going in the wrong direction." Are you prepared for this walk?

Introduction

This book covers what scientists know about space, time and matter. It also answers theologians' belief that God is the Alpha and the Omega and is omnipresent.

I believe this makes the book unique and different.

The fullest interpretation and complete understanding of the universe can only be derived from the broadest, richest and widest application of ecology—a *robust ecology*—a new ecology finally freed from its traditional limitations as merely interesting branches of biology or sociology.

The rock we live on that we named Earth had its beginning from the Big Bang, as did everything in the universe. The Earth formed about 4.5 billion years ago as a red-hot mass within our solar system. As the Earth cooled down, the atmosphere and oceans developed which provided an environment for the first living one cell organism. The chapters that follow provide the science and ecology of it all. The ecology of it all is, and must be, in harmony.

When I think that God (omnipresent) is in every atom of my body, I get euphoric and harmonic chills. We are one. God is not "other." There is no He or I—we are one. When I lie down to sleep at night I experience a special and beautiful sensation. You probably have your own way of reaching this feeling. I review the day and ask myself if I truly accomplished anything. Did I behave during the day in ways that would please Him? Only honest thoughts are allowed because He knows all.

All elements within the universe abide by the laws of nature and what the scientists know.

However, things changed when Homo sapiens arrived

on the scene. Ever since we appeared on Earth, we humans have been under attack by viruses, bacteria, and horrific weather. We have also been under attack by our fellow humans.

The interwoven interrelationships of all ecosystems are just like the fabric of the time-space continuum that I describe later in this book. There is no distinct separation or difference of either one from the other. Stated more succinctly, when we compare all interrelated ecosystems and the time-space continuum, there simply is no difference among them—because they are a woven fabric of oneness.

I am going to write about how we know. I will discuss scientific facts and theories and theological facts and theories. The domains of science and theology are fluid, not fixed. Even when you believe you have arrived at a fixed conclusion, always remain skeptical. This is the ecosystem of knowledge itself, and we are in it for this journey. Knowledge's ecosystem is not fragile, but robust. All appearances are illusions, and nothing is forever. The biggest envelope contains choice as well as knowledge itself; the biggest "container" includes everything: God, the universe, and all living things within.

Ecology appears eclectic, a hodge-podge, until we uncover and see the connectedness behind the scenes.
For the past several years, my mind felt like a disordered collage of entropic ideas about science and the Almighty. I took on the mental task of developing an ecology of the jumbled ideas. Many ideas (and "stuff") have their own ecosystems, but was it possible that they all are one?

This book is my thinking odyssey—my challenging journey to discover and know the final result: *All seemingly unrelated ideas and things are one.*

No longer is there a mish mash collage, but a beautiful and coherent picture with all the colors serenely stroked as only the Almighty could do.

Like the philosophers of old, I derive great joy from pure, uninterrupted thinking. I urge and invite you to take this journey. Along the way, form your own ideas, thoughts, and conclusions. Hopefully you will also enjoy yourself.

For centuries, theologians and scientists have been at odds with each other. There was and is no need for that.

I dispel the argument of creation versus evolution as moot.

In the coming pages we will briefly cover the creation of the universe; geology; biology; the Earth's evolution; physics; genetics, cloning and religion—all within the realm of ecology.

No longer will you view ecology on the dark side— acid rain, ozone layers. You will come to see the ultimate harmony of our ecological world.

The book is relatively short, but like a good nutritious dinner, there will be much to chew on:

I also theorize that *the universe is a perfect perpetual entity.* All matter, time and space, *including God* (whom I have added and is at the controls) were created with one giant explosion of a small mass called the Big Bang theory. The universe is still expanding—this fact is backed up by scientific measurement. When the universe stops expanding, it will begin to collapse (like a dying star becomes a Black Hole) and once again will become a small mass. It will then explode and once again create matter, space, time *and God.* Once again, I have added God as part of the Big Bang creation. God remains ever-present and omnipresent.

What makes us humans unique is our capacity to think, and in many ways we have abused that incomparable power. Who knows how far our minds can go?

Philosophers contemplate whether man is basically good or evil. Judging from our negative impact on the earth's harmonious ecology, so far we have defined ourselves as evil. How did we lose the good that humans were given?

As Albert Einstein said, "We cannot solve our problems with the same thinking we used when we created them." Living creatures less intelligent than we are have survived through physical evolution. Time is running out for humanity.

Our last best hope for survival and harmony is a rebirth of our species through thinking.

How I redefine and seek self-determination from writing this book is very personal, as it should be for you on this journey.

CHAPTER 1

HUMANITY AND THE UNIVERSE

When I was studying the Mandarin Chinese language at an adult class, the instructor presented a one-night session in Chinese Art.

Using a projector, she showed many slides of landscape photos of huge mountains, blue skies, rivers and usually a very small painted human, almost indistinct in the picture. She explained that the small scale of the person was to express visually how insignificant humans are in the universe. The painter was making a philosophic statement.

With an open mind, I have been studying humans and the universe for about five decades. I still have many questions. When we deal with the abstraction of a beginning and an end, does God also have a similar existence? In Revelation it says, "I am the alpha and the omega (first and the last, the beginning and the end), says the Lord God." This literally states that life is not eternal. Eternity has no beginning or end. We will never know. We cannot think in this frequency or wavelength. Our minds, like a radio, can pick up limited frequencies. In a somewhat reverse thinking process, Socrates' epistemology (philosophy of knowledge) states that we were born knowing everything, and we spend our lives remembering what we already know. We can't understand "end" or eternity, and its realm, because our life experience can't access an end.

To think about an end of life without a hereafter is, and has been, unquestionable for most religions. Yes, indeed, most believe that life is eternal. This fits well with my theory that our universe is eternal (one Big Bang after another). I further propose that there is a oneness of the ecology of the universe. But if it is eternal, does it have a shape? If it has a shape, how can that be since space is created by the Big Bang? Most theorists in cosmology believe that the universe is flat with only a 0.4% margin of error. The model is based on observational data and not mathematics. To this point, I ask what lies around the outside of a flat universe with no space? Some cosmologists believe there is more than one universe and that there are countless universes. One theory just creates more questions.

Now back to humanity.

About 5,000 years ago, Egyptian pharaohs were mummified and buried in huge pyramids. Domestic helpers and other personnel were executed and buried with the dead pharaohs to serve him in the afterlife. Food and riches were also placed around the sarcophagus.

On the other side of the world (210-209 BC) in Xiam, China, Emperor Qin Shi Huang had a burial ground made to house 8,000 terracotta soldiers, 130 chariots and 520 horses. They would protect the Emperor in his afterlife. The site also included non-military figures: officials, acrobats, strongmen, and musicians. The Emperor would thus have people to entertain him and to rule over.

In both cultures no mention has been found of a divine revelation absolutely promising an afterlife. The ancients, like modern man, *wished* there were an afterlife. Why are we humans so obsessed with a hereafter?

The early Egyptians and Chinese believed in a hereafter and yet did not develop or possess religions that endured to modern times. Did they have a hard, strong "wish" or longing that there be one? Today's religions have transformed the wish to a certainty. But the nagging fact remains even to this day: no one knows for sure no matter how you dress it. It is a belief, not a reality.

In the timescale of the universe, we are newcomers. It is widely believed by astronomers and geologists that the Earth is 4.5 billion years old. By reconstruction of the Bible, the Earth is only 6,000 years old, and (Genesis 5:5) Adam lived for 930 years or 16 percent of it. The Pyramids of Giza are estimated to be roughly 5,000 years old and artifacts dug up by geologists and archeologists are in the million-year-old category. By use of carbon 14 dating, the artifacts placed in the rock layers are scientifically reliable dating measures.

The theory the Big Bang, as the event that created everything in the Universe, is widely held by most scientists. By estimating, the time elapsed since the Big Bang would place the Universe at approximately 13.75 billion years old. The size of the Universe is estimated at 93 billion light years. A light year is a unit of length equal to 6 trillion miles. Their research is supported by a number of scientific research projects such as microwave background radiation measurements by Wilkinson Microwave Aniso-tropy Probe and other probes. Additionally, the Universe is still expanding at an accelerated rate.

In the mid-1960s, Bell Laboratories in New Jersey built a 20-foot horn antenna that could pick up noise in outer space. When they first tested it, a constant sound was detected in all directions, day and night. Suspecting the constant noise was due to some flaw in the antenna, the scientists cleaned it up by removing all foreign matter from surface. When it was retested, the results were the same. Bell Labs conferred with scientists from Princeton and M.I.T.

Bell's was amazed when told that the noise was the answer their colleagues were looking for in order to have concrete evidence of the Big Bang theory. The constant "sound" was the radiation left over from the Big Bang.

While working on my MBA at Fordham University, I had a brilliant woman friend who worked at Bell Labs. We would discuss the Big Bang and other theories, various topics and unclassified projects topics after class.

What is the Big Bang? I will present a brief abbreviated explanation, one that in its entirety is the prevailing cosmological model that explains the early development of the Universe.

Very accurate and powerful instruments revealed that the Universe is expanding in all directions at great speeds. Among other evidence, this proved that the Universe must have started at one point and from an incalculably dense, but small mass. Before this mass, or Big Bang, the Universe had no matter, space, or time which was all created at once. This is what they call the singularity theory.

Now the guesswork begins. Where was God? The life of the universe can be hypothesized that the acceleration will end and the work of gravitation pull will reverse itself and once again collapse into one extremely dense mass. It then will start all over again, truly a perpetual universe with no end. But all our knowledge of the physical world does not answer the existence of God. In our mind, like the radio frequencies, it is out of our range of thinking. Does God have a size?

Let's look at the smallest item in the Universe, the atom, or so we thought for centuries. The study of the atom is called quantum mechanics. Beginning with Einstein, the reconciliation of the laws of the atom and those of the Universe has baffled the scientific world. They gave up and pronounced that it could not work. But research still goes forward.

The insides of an atom are almost totally empty. There are seven billion, billion, billion atoms in the human body. That's 7 with 27 zeros after it. The extreme speed of the electrons around the nucleons gives the illusion of matter, but that is all they are. They are almost indestructible. Only by fission and fusion can they breakup, and when they do, it's with a great instantaneous explosion like the atomic bomb (fission—split the atom). On the sun's surface fusion is constantly going—creating the tremendous heat and light. Scientists are working diligently on creating cold fusion by that will release more energy of a fused atom than the amount of energy to create it. No success so far.

Atoms continue their lives for eons, but reemerge into different matter. For example, you may have atoms in your body from a dinosaur that lived 200 million years ago. And yes, they have God in them.

Considering that we are mostly air, as are all the things around us, is it possible that after one million attempts we could walk through a wall? Not probable, however.

The most advanced research of the atom is called the String Theory. In short, the String Theory posits the idea that what little there is of the atom is merely waves. It is, however, a contender for the Theory Of Everything (ToE) and could possibly reconcile quantum mechanics (the study of the atom) and the general relativity of the universe.

The physical world is truly complex. I remember seeing a movie depicting Sir Newton in his laboratory. He had just completed building a small mechanical model of our solar system. When he turned a hand-operated wheel, the moon revolved around the Earth, the Earth rotated on its axis, and all the planets revolved around the Sun. A colleague entered his lab and asked Sir Isaac what he was working on. Newton demonstrated the device to the colleague who was astonished about the invention. He inquired, "How did you make it?" Sir Newton said, "It all came together by itself."

Did our Universe just by happenstance come together? I think not.

Was this another masterful work of God?

Since the Universe is so vast, scientists have chosen light years as the basis to measure distances. Light travels at the speed of 186,287.5 miles per second. Multiplying the amount of seconds in a year: 31,557,600 times 186,287.5 miles equals 5,878,786,100,000 miles or approximately 6 trillion miles in a year.

The relationship between time and light is worth understanding. Time slows down the faster you travel. Even the satellites around the Earth that are used to operate your GPS gadgetry have to adjust (albeit very small) their clocks to account for their speed. If they didn't, you may turn at the wrong exit or street.

An extreme fact about the speed of light is that once reached, time stands still and if you were to surpass it, you would go backwards in time. Nothing, however, goes faster than light.

When you look at a clear night sky, you see millions of stars (suns). Since the distance of most of the stars is so far away, you may just be seeing the light from a star that died millions of years ago—once again, the illustration of relativity.

I'm beginning to see a Master at work here and I doubt we look like Him.

Let's briefly explore Einstein's famous formula of general relativity $E = mc^2$ since it incorporates the speed of light. $E = mc^2$, or energy, equals mass times the speed of light times the speed of light. In our language it means that when mass, any tangible item (stone, Earth, you) reaches the speed of light (186,287.5 miles a second) times itself or approximately 33 trillion miles a second, it converts to energy.

Who wrote the book of Genesis? No one knows. Some say Moses, but he lived 2,500 years later. Were Adam and Eve literate and if so, what language and on what did they write on? Genesis says that we were created in His image. I find that a bit egotistic and beyond any of the laws of science. We know since grade school that stories change vastly when passed on from student to student. Tell the first student a short verse and whisper it around the class, and the end result is totally different. The ages of just four people in Genesis: Methuselah 969 (Genesis 5:27); Adam 930 (Genesis 5:5); Jared 962 (Genesis 5:70); and Noah 950 (Genesis 9:29).

If just one adult who has never been exposed to religion read Genesis—see his or her reaction. How many adults have read the entire Bible while at their current age, and all in less than one year?

Who was the second woman in the bible and where did she come from? Don't throw my book away yet. I do believe in a Supreme Being, but there are some unsettling questions in my mind.

If you feel uncomfortable with these factual questions or disdain for me, ask yourself why. Think it through. Can you put it into writing and convince yourself that these questions are uncalled for?

I love the evolutionary process and how both animal and the vegetation world have flourished over the centuries. But notwithstanding, that I wonder how two sexes occurred. Some species in the animal world are asexual but they are rare.

Modern man can only live on a planet like Earth. One small change in the atmosphere, or basic minerals in our bodies, or the size of the Earth and its associated gravitational pull, could destroy our species.

Some scientists believe that without the moon we would not be here. We know that the moon's gravitational forces cause the ocean tides. According to Dr. Fred Gortz, who worked as a scientist, researcher, and teacher, the moon is good for a whole lot more than simply lighting up the night sky. If you wonder what your life would be like without it, the surprising answer is that you wouldn't be here to wonder at all.

If we lived on the planet Jupiter, which is biologically impossible, our legs alone would have to be the size of tree trunks since the gravitational forces, because of the planet's mass, would be so powerful.

We are made for this planet only and I believe it was by way of evolution. Our closest living relative is the Chimpanzee, with 98% of the same gene makeup as ourselves. I believe that God created us by evolution and not in the Garden of Eden. Really, what difference does it make with so many other uncertainties that we have about God that will never be answered while we are alive on Earth?

When I was a youngster I remember reading a science fiction piece about going back in time. Today, I believe it is called the Butterfly Effect. In a high tech laboratory, they invented a chamber/capsule that could hold a man to go back in time. They found a brave candidate willing to try the experiment. The chamber had a strong glass door so that the adventuresome person could observe the habitat once the chamber landed millions of years back in time.

They warned the man not to step outside the chamber so not to destroy or affect a change of anything. He promised them that he would remain inside the chamber/capsule. When the capsule landed in plush, green grass with scattered trees and bushes, the man was astonished by the savannah's beauty. After a few moments of awe, he just had to step out for a better look all around the capsule. When doing so, he accidently stepped on a butterfly which now would flutter no more. He quickly re-entered the capsule and within minutes was swished back to the future he left. He landed on barren land. Everything had changed.

The point of the story is that even a small, delicate change millions of years ago would have a multiplier effect that would drastically affect the future. Just think about how extraordinary it is that you are here. This is not a theory but a truism. It is integral to realize as you read further into this book.

Two gentlemen, from different backgrounds, relate to the "Butterfly Effect," and I must briefly mention them.

Sir Isaac Newton's (1642-1727) third law of motion states, "When two bodies interact by exerting force on each other, these forces (termed the action and the reaction) are equal in magnitude, but opposite in direction." Jim Rohn (1930-2009) had humble beginnings but became a legendary speaker and writer in personal development and human potential. One of his best speeches was on a subject that was very close to his heart—"Everything Affects Everything Else"—or, the Butterfly Effect.

My book is based on these ideas, except it is on a grander scale—and includes the universe and God with our knowledge of the sciences leading the way.

This leads me to the belief that God does not tinker with anything on Earth. The multiplier consequences, as mentioned above, would make us more like his puppets rather than be fully accountable for our actions. I'm sure that the 20 million Russians killed in the Second World War prayed for their lives, as did all the Jews in the concentration camps. How many prayers can you prove were answered by God, and not really *post hoc ergo propter hoc* (*after this, therefore because of this;* a fallacy "based upon the mistaken notion that simply because one thing happens after another, the first event was a cause of the second event." (Source: www.skepdic.com/posthoc.html) There is also the placebo effect that affects change, but it's only in your mind.

Throughout our existence we have prayed to God for peace, love, help, etc., but up to this very moment we continue with our past behavior. Do we want God to make us be in peace? That is out job. There is no evidence that we will not continue life as usual. Why should He change us—are we not accountable for our own actions?

Another way of thinking about God's lack of interference is the flip side of the above. Many people blame God for all the catastrophes that occur and cost many lives, but hurricanes, tornadoes, and floods are just natural events that have been going on since the Earth's creation. Many refer to these events as "acts of God."

For most of my life, I prayed like most Christians— once again, just following the crowd. But I began to think about how God thought about us, not just now, but ever since He created us and the Earth—all of the wars, murders, and other atrocities throughout our history. How did we become so immune to wars? He gave us a wonderful body and fantastic Earth, and we repay Him with this kind of behavior, and yet even pray for more help.

Therefore, I talk to God each night and tell him how sorry I am for what we have done and how I could only imagine how disappointed He must be in us.

Now that I have developed the theory that God is in every atom of my body, praying to God would be like praying to yourself.

My theory is just that! My motives are pure—bring unity in its biggest form to all humankind—God. Although we may see Him through different lenses, there is just one God. The facts are not disputable—we have not pleased our God in our historical behavior.

It is not to say that our world has not had its ample share of wonderful people through the ages—those who are by their very nature pure at heart. They never expect to be rewarded for the love and kindness they have given others. My greatest pleasure in life has been to see humans at their best. They have not gone unnoticed.

Man is infinitely tiny in the universe. The universe, according to astrophysics, is comprised of much more than the suns (stars), planets, meteors, comets, and black holes. The universe also is dark matter and dark energy. According to "The A Register," all items that we can see, even with the strongest telescopes, total only between 4 and 5 percent of our universe's mass. The universe is vastly more complicated than our solar system, which I described earlier.

For us to find Earthlings like us would require a solar system exactly like ours. If we got a signal from some distant planet, a signal comprehensible to us, then we would have found confirmation that a life like ours is possible.

Lastly, I will mention briefly and very simplistically the two-dimensional space-time theory. Euclid described that our universe has three dimensions of space, and one dimension of time. Recently, astrophysicists believe that space-time is woven into one fabric called the space-time continuum.

Our universe is indeed a very complex place to live in. I will explain later how I believe that God must be a part of it all.

CHAPTER 2

HUMANITY'S EARLY HISTORY

Did we evolve on the Earth as Darwin, archeologists, anthropologists, and other scientists believe, or were we just created in the Garden of Eden? To me, both were by the hand of God. If I had to make a choice, I would agree with the Darwinians because of the enormous scientific evidence. Or were we from Eden, based on the Book of Genesis, and whose author is unknown?

Let's go back to the time when the Book of Genesis and the Garden of Eden took place. For the purpose of my book, many Bible scholars agree the event occurred approximately 5 to 6 thousand years ago.

What was going on around the world at this time? My research below is primarily based on *Hammond Past Worlds: The Times Atlas of Archeology* (1988).

In the new world, the "America's," there is evidence of potatoes in Peru (6,000 BC); cultivation of maize in Mexico (5,000 BC); pottery in Columbia and Ecuador (4,000 BC), and fishing nets and the cultivation of cotton in Peru (3,500 BC).

In Europe, the first wheeled vehicles appeared in Hungary around 3,200 BC; the beginning of the full European Bronze Age began about 2,300 BC; the main phase of Stonehenge about 2,000 BC, and Cretan hieroglyphic writing about 1,900 BC.

In Africa, Egypt's first walled town was built in 3,400 BC; the first evidence of Egyptian hieroglyphics is dated 3,000 BC, and the first Egyptian pyramid built at Saqqara about 2,700 BC.

In West Asia and South Asia the earliest writing in the

world from Tell Brak, northern Mesopotamia (3,250 BC); Cuneiform script in Mesopotamia (3,100 BC); use of the plough in the Indus Valley (2,600 BC), and the development of urban civilization on the Indus Plain (2,500 BC).

In East Asia, the first evidence of agriculture in Korea, millet cultivation (3,000 BC); silk weaving in China (2,700 BC), and early bronze metallurgy in South-East Asia (2,500 BC).

"According to the Bible, the Garden of Eden was located somewhere in southern Iraq where the Euphrates and Hiddebel (Tigris) Rivers flowed into the head of the Persian Gulf—that is, they flowed on a modern landscape. This is still recognizable today." (From The American Scientific Affiliation, March 2000)

The names *Australopithecus* and Lucy are now familiar to many. Just a few years ago the scientific community was truly shocked with the discovery of "Ardi," or *ardipithecus ramidus.*

Joel Achenback of the *Washington Post* reported that "Ardi lived 4.4 million years ago in the woodlands of East Africa. She stood about 4 feet tall, weighed 110 pounds, and had long arms and short legs." Yohannes Haile-Selassie, the paleontologist who found the first two bones of Ardi in 1994, is quoted as saying the discovery of Ardi "further confirms that Ethiopia is the cradle of humankind."

However, scientists have unearthed several species of human-like remains that are up to 5 million years old. In searching for the "missing link" of humans and the ape family, they have uncovered Java man, Peking man, Cro-Magnon, and Neanderthal. How much evidence do we need before we will accept evolution as our Creator's way of bringing us into existence?

Now that we have some information about other

parts of the world when Genesis was written and Adam and Eve were created, let us bring our minds forward to the appropriate time Jesus the Christ was on the Earth:

In Europe, Augustus became the sole ruler of the Roman Empire (27 BC); Roman pottery traded to northern Europe and southern India (1 AD); in Africa, the foundation of the Roman colony of Carthage was begun (46 AD); in West Asia, extensive irrigation schemes in Ceylon occurred (1 AD); in East Asia, Chinese silk traded to the Romans (50 BC); and in the New World, Moche, famous for their gold and pottery, dominated the north Peruvian Coast (1 AD).

Why is there such a disparity between the Bible and scientific discovery? Why did God choose Iraq and such a short time ago, according to the Bible, as the place to inhabit the Earth with humans? Why did He speak to so few people and in one small section of Earth? Why hasn't He spoken to anyone in such a long time?

Why would He expect us to believe in a handful of people that He supposedly spoke with? Why wouldn't He give us all an equal footing to know Him and, as we do as parents, guide us along the way? Why doesn't He intervene when a vast amount of people are wrongfully led by a false prophet? These are not sacrilegious questions to a true God-based religion. And why do we feel guilty when we ask ourselves these questions?

Do scientists feel guilty when they discover irrefutable evidence that contradicts their religion? Galileo was almost excommunicated when he improved the telescope and realized that the Earth revolved around the Sun and not the other way around, as the church taught.

If you hold yourself accountable for the entirety of your well thought out beliefs, it does not come easy. The real paradox is this: If we conduct our lives honorably, grow educationally, break no laws, but those around us ostracize us, what should we do? Should you be yourself, or what others want you to be—that is the question.

It certainly would be a difficult task to form a governing body that permits all citizens to act as individuals. Our democracy tried to do so by permitting many freedoms and for a long time it worked well. But slowly and surely we are losing our individual freedoms, and at the same time many of them are being abused. Are we truly a democracy, or has socialism and dictatorial rule entered our system of government?

Henry Ford once said that you could have any color car you want as long as it is black. Our government was originally written so that all men were created equal as long as you were white and a male. This leads to the question: Is humanity capable of being sustainably governed?

Great empires of the past came to an end mainly because of irresponsible leadership and the social decay of the virtues and morals of the governed population. They then became vulnerable to outside enemies and perished.

It is thought provoking to compare the Russian and the Chinese governments over the past three decades. Previously, both were fully Communist countries. China's economy has since flourished since they became a hybrid Communist-Capitalist society. The Chinese people are now reaping the benefits of free enterprise, but they still maintain a strong hold on China's governance with a politically Communist rule. On the other hand, Russia changed to a democratic society overnight and created chaos and a vacuum for economic corruption. I was there with the US Secretary of Labor when it occurred.

The reason China's change worked so well is that people were permitted to quickly improve their financial position, while the leadership kept a firm hold on the reins of governance.

I believe that our country has entered into a slow, but certain, economic downward slope without an end in sight. I should not voice my opinion without a solution to the problem, but it needs to be recognized and we all should adjust ourselves for the inevitable. I will commit myself to one more dilemma: What person, who has all the intellectual requirements, talent and skill, and sincere individual qualities, would truly want the job of President of the United States today?

Groucho Marx once said, "I would never join a club that would let someone like me in." Personally, if I were fully qualified to become President, I would not take the job because it is a job with problems I know I could not fix.

The world has come a long way in the past century, but are we any happier? I implore you to think where we will be in the next hundred years. Will God say to Himself, I have seen enough decay in the human race and it's time for Armageddon (Revelation 16:16)?

CHAPTER 3

THINKING AND EXPERIENCING

When I reflect on my teenage years, I now see that most of what I was told became what I believed. My schooling in high school was based on remembering (history, English, etc.), not *thinking*. An exception was "word problems" in math. Applied arithmetic required thinking and was a real challenge to many students.

In college, self-disciplined hard thinking is mandatory, and may be the main reason so many freshmen dropped out. Some were there just to party.

Eastern religions (Buddhism, Taoism, Confucianism) teach you to free yourself of "wants." You may have heard the saying, "Don't wish for something, because you may get it." I found through personal experience that both tenets are true. Another accurate saying is "money can't buy happiness"—unless you don't have any. But if you acquire a vast amount of money and become *nouveau riche*, you may find out that it is true. What does the Bible say about this? "Again I tell you, it is easier for a camel to go through the eye of a needle than for a rich man to enter the kingdom of God." (Matthew 19:27)

On one of my numerous trips to Japan, I became aware of a certain custom of thinking. In Japan, when a subordinate goes to his boss's office and sees him busily moving papers and writing, he feels comfortable interrupting him. But when he sees the boss looking out a window, he does not disturb him because he knows the boss is performing an even more important task. The boss is doing his most serious and important work. *The boss is thinking.*

Augustus Rodin, the great sculptor from France, believed thinking was important enough to sculpt his famous piece, "The Thinker."

I believe that God is going to hold us accountable for the lives we lead, and will judge every act we do while on Earth. Are our acts well thought out? If you were to take inventory of all your decisions, would you have made better choices? The old adage, "look before you leap," perhaps should be "ponder before you plunge." Every thought you put in action will have subsequent effects—not just on you, but also on myriad future events. The Butterfly Effect is well played out in the popular 1946 movie, "It's a Wonderful Life."

Earlier I mentioned that "word problems" in math were rather difficult. Below I am going to present you with a word problem to solve that requires thinking ecologically about the Sun and Earth. You may find it fun to try on a friend. The only thing you need is a 12-inch ruler. Ask them if they can tell how high a telephone pole, or something similar, is without measuring or touching it. No, they can't climb the pole or call the phone company or Google it or get any other outside help.

Answer: place the 12-inch ruler straight up along side the pole and close to the pole's shadow. Measure the ruler's shadow. For simplicity purposes, let's say the ruler's shadow is 6 inches or one-half of the 12-inch ruler. Then measure the pole's shadow; let's say it is 30 feet. We know the ruler's shadow is one-half of its height. Therefore, the pole must be twice its shadow, or 60 feet. I am surprised so many people I pose this problem to cannot figure it out.

Grenville Kleiser said, "To every problem there is already a solution, whether you know it or not." One of my great joys in life is discovering the truths that were *always and already there.* "There is nothing new under the sun." (Ecclesiastes 1:9)

IBM created its "THINK" motto several decades ago. As CEO of a global corporation, I had Grenville Kleiser's saying framed and hanging on my office wall. On several occasions, when one of my executives with a problem would come into my office, I would point to the quote. The best way to form a good company culture is to have all employees realize that they are valuable to the overall performance of the company. I wanted the quote to be part of my company's culture.

Like the rest of our body, our brain requires a lot of exercise to function well. "Use it or lose it" may not fit into your way of life, but it should. You may be a genius, or nearly one, but you likely never used your full potential of thinking power. Does the size of our brain matter? No, there is no correlation between brain size and intelligence. According to *Science Now* magazine (sciencemag.org), a team of scientists concluded that "one parameter that did not explain Einstein's mental prowess ... was the size of his brain. At 1,230 grams, it fell at the low end of average for modern humans." Human brains have an average weight of 1,320 grams (about 2.9 pounds).

According to Dennis E. Coates, PhD, "The part of the brain that produces the most striking differences in personality is the cortex, the outermost layer of brain matter. This is the thinking part of the brain, where perception, language, learning, planning, problem solving, and most high-level functions are processed." Where does the nearly- empty atom fit into the structure of the cortex? The hierarchy goes from sub- atomic particles to atoms, then to molecules (which make up cells), then to tissue, and finally we reach the part of the brain called the cortex. The molecules contain DNA, RNA, protein, etc. So our brain is actually fairly empty as is our entire body and the universe. So when someone says, "You're all hot air," they are literally and scientifically fairly correct. But it is a stretch when someone says you have "a pea brain." Brain size does not correlate with mental ability.

What is the empty space of the universe? Quoting from NASA science, "More is unknown than is known. We know how much dark energy there is because we know how it affects the universe's expansion. Other than that, it is a complete mystery. But it is an important mystery. It turns out that roughly 70 percent of the universe is dark energy. Dark matter makes up about 25 percent. The rest— everything on Earth, everything ever observed with all of our instruments, all normal matter—adds up to less than 5 percent of the universe. Come to think of it, maybe it shouldn't be called 'normal' matter at all, since it is such a small fraction of the universe."

What do you think? This is a "thinking" chapter. Is it possible that God is in all of us, including our bodies? He is not an individual entity, but part of everything.

In Christianity, God is believed to be omnipotent (can do anything), omniscient (knows everything), and omni- present (is everywhere). Could the physical "emptiness" in all of us be the spirit of God? But He is not the decision maker—you are; otherwise you could not be accountable for the life you live.

This thinking (and image) of *God-in-us* is an abstract- tion beyond our limits, like a radio frequency we cannot tune into. This is why faith and belief function to fulfill our desire for a Hereafter. We demand an answer to the question, "Why must we all die?" We die because organic matter ages with time.

Beliefs have a tendency to become a fact when culti- vated over a long period of time. I knew a senior citizen who thought the USA's landing on the moon was a hoax, and there was no way anyone could change his belief.

Do dogs think? If you have owned one for a number of years, you would probably agree that Fido could think. If you have any doubts, just surf the Internet. There are countless articles confirming your pet's thinking skills.

My personal experience goes back to my living in Bennington, Vermont. We owned a female Siberian husky named Cheyenne. She was in her element with the cold and snowy winter months. I could write a book just on our experiences with Cheyenne.

On the weekends when I was home, I would take her for a ride to the top of a snow-covered mountainous hiking area. All I had to say was "car" and she would run wildly back and forth to the front door. Her tail would wag nonstop and she would howl and bark at the top of her lungs. She could "understand" the word?

In the hiking area, I would remove Cheyenne's leash and we would wander for a couple of hours. On one outing, we needed to cross a stream that was thinly covered with ice. She sensed the ice would not hold her weight, so she refused to cross. I could hear the faint sound of running water below the ice. We both then walked along the stream until we reached a narrow passage. I sorely miss those wonderful hikes.

When we were not hiking, I would take Cheyenne for walks around the neighborhood. The streets were lined with antebellum homes (my house was built in 1821) on large grounds. Most often the neighbors were inside their houses. Cheyenne knew the neighborhood by heart. On one occasion a lawn mower had been left outside a home. Cheyenne knew it was an object out of place and barked at it until I calmed her down.

These are two distinct thinking patterns in dogs. The

first is referred to as "word" recognition, and the second is "non-verbal."

Most dog owners will relate to these incidents, and probably have just as many as I do.

Reminiscing about Cheyenne reminds me again of my backyard ecology. As I was recently looking out over my yard, I spotted a Grey Hawk perched on top of the fence. I usually see hawks in flight, but this was a rare sighting for me because of the hawk's motionless state. It was totally still except for occasional slight head movements. After about 10 minutes, the hawk suddenly swooped across the lawn and captured an unsuspecting small bird. The bird had been midway up a tree, and was probably eating insects. Last year harvester ants denuded the tree. The ecology of hawk eats small bird, small bird eats insects, insects eat tree, and tree suffers less harm, is fully at work.

I wonder what the Butterfly Effect will be centuries from now. Just think how incredible it is that we are both here today and how our lives will impact the future.

Take a look at Thomas Edison in this context. He holds the record for patents at 1,093. What a fabulous mind, perhaps the epitome of inventive and creative thinking.

Having a great mind does not mean that gifted people also have good common sense. I remember reading about John Stuart Mill (1806-1873) who was called "the most influential English-speaking philosopher of the nineteenth century." He was also a Member of Parliament. But, on occasion, Mill could not find his way home.

When my children were growing up, I tried to convince them that a strong ability to communicate both in writing and speaking was critical for their future.

I told them, for instance, that a stone might be very intelligent, but not noted because it could not communicate. A bit far fetched, but you get the point.

If you are a person who has rarely used the creativity inside your brain, you are missing many opportunities. This great latent asset needs to be awakened. Don't ignore ideas you may have because you are afraid you may fail. I heard a story about Edison that I have incorporated into my way of thinking. When Edison was at the peak of his inventive pursuits, a person asked him if he ever failed and Edison replied, "Yes, and I learned something—it [my experiment] doesn't work." Failing is part of the natural creative process. Failing is superb feedback—or *communication.*

We all have met people that constantly say they will "try it someday" and never follow through. Use this simple demonstration when they mention it again. Ask them to "try" to throw a pencil. When they toss the pencil across the room, tell them they could not just "try" it because the pencil actually goes or does not go across the room. In short, either you do it or do not do it. Trying is not an option.

In my Memoir (2009), I wrote that "my mind is my playground, and now I see it is cluttered with incongruent fixations. It needs cleansing, a rebirth to once again be a happy place to visit."

For the past four years I had to do a lot of re-thinking, including a close look at what makes me happy. At first I was surprised that my answer was "accomplishments," and now I am convinced. From experience, I can honestly say it is not being rich.

In mid-afternoon I can turn on some soft music, lie on my couch, and happily ponder whatever is on my mind.

One clearly productive and beneficial realization about our mind is that you can usually only be in one state of mind at a time: happy, sad, scared, and tranquil, loving, hating, etc. Knowing this, I try to manipulate my mind into the desired state. You can think yourself into serenity. *God grant me the serenity to accept the things I cannot change, courage to change the things I can, and the wisdom to know the difference.*

Another inspiring realization can be found in Eckhart Tolle's writings and seminars regarding the "power of now." He teaches us how to live in the now. No watches, no clocks—time simply does not exist. The past is gone and buried, and the future has not happened. You live in the present and the present only. You can throw away your worries about the future because the future does not yet exist. Enjoy each moment. This is real living. You only have one state of mind to live in, so make it in the present.

You will read more about thinking in the last chapter, "Your Unopened Gift."

CHAPTER 4

YOUR WORLD

"How fast are you going?" I ask.

You reply, "Don, I'm sitting still with my cup of tea which is also motionless." I reply, "It would appear so, but it's just like the illusion of your appearance. You are nearly all air. Recall the empty-ness of the atom? You are made of nearly-empty atoms."

And you are speeding through space in many directions. The Earth is rotating on its axis, and the Earth is revolving around the sun. The sun orbits the center of the Milky Way galaxy, and the Milky Way in the local group of galaxies. The local group falls towards the Virgo cluster of galaxies and all speed out from the explosion of the Bing Bang.

Next question is about you. What is your makeup and individual ecology? There are three of 'you": the person you think you are, your personality that other people see, the person that you really are. The closer these three are redundant or similar, the saner you are. Then there is: you in your "time and space" slots. The Big Bang Theory states that time and space were created at the same point of the explosion. Then there is the physical you and God (omnipresent/everywhere). It is getting a little crowded in you.

Since we live in our empirical world, we just believe in it rather then accept or try to understand all the science behind it.

I can imagine a sci-fi movie in which a research laboratory creates a hand-held device that when pointed and triggered, causes all the electrons in a defined area to stop spinning. The atoms die—not from fissure or fusion.

The target area becomes invisible. In the wrong hands, this hand-held device could be the ultimate weapon.

Perhaps another movie could be made where a woman has a hand-held device that aligns her own atoms and the atoms of a wall, which she could quickly step through. My sister Carol loves this idea because she likes to believe in ghosts. Halloween is her favorite holiday.

Your mind could also be a fun place to visit. All you need is a comfortable chair or couch and free time alone. Clear your mind of worrisome thoughts and choose memories that give you the most pleasure. Think about exotic places to visit. These mind-visits could be better than the real thing— no bugs, no fees, no bad weather, and disease-free. Remember, whatever you are thinking about will create that state of mind (happy or sad). There is no limit to where you can take your mind, or your mind can take you. You can think about inventing something, ways to improve your life, doing a good deed, writing a poem, and so on. Take inventory of the activities that give you the most happiness and pursue them. You can do what I am currently doing outside the box: challenging all my beliefs (e.g., politics, meaning of life, religion, and prejudices). This is the map I have chosen to seek my individuality and self-actualization. Using your brain can also be healthy for you. Use it, or lose it.

How is your memory? Would you like to impress people with a mental exercise that I learned at a Dale Carnegie course? You must first memorize twenty commonplace items and in numerical order—items like a chair or a Christmas tree or a cup. The chair could be number one, the tree number two, etc. Then have someone slowly name twenty items that you must remember in the order given. Now visualize the item as if it were placed on a chair, or on the table, or hanging on the Christmas tree, etc.

Commit it to memory—take a minute or two. When they are finished, ask them to name a number. If they say four and it was a candle, which you mentally placed in the cup, you say candle.

A century ago, the average human brain had very little thinking to do in everyday, routine life. Then came the tech revolution: cars, planes, phones, radio, TV, microwaves, computers, NASA, genetic engineering, et al.

Some inventions, like computer-related devices, are growing exponentially and most come with long and complicated operating manuals (in print or more commonly online). What is amazing is that the brain was equipped to handle it all before hand. Its pre-designed capability is not like the evolution of the species. It didn't adapt—it was simply there ready and waiting for the challenge. This is rather unusual in nature—to have a brain ready for a brand new task. However, some gifted humans thousands of years ago made full use of their brain. This is the second enigma (the first is opposite sexes) that still baffles me. God, not evolution, created mankind. Or, God at least created these human characteristics. In any case, I believe that God created everything. He just used evolution as the process by which we are here.

This is not the 1925 Scope's trial, but my reconciliation of the two opposing sides. In the Scopes' trial, formally known as *The State of Tennessee v. John Thomas Scopes*, the State accused Scopes of violating Tennessee's Butler Act that made it unlawful to teach evolution in any State-funded school. This was clearly the power of government encouraging ignorance. Scopes was found guilty, but the verdict was overturned on a technicality and Scopes went free.

How does all this affect you? Actually, it works the other way: How do you affect the world?

In the classic movie, *It's a Wonderful Life,* an angel shows James Stewart how worse off his city would be if Stewart had never been born. A host of citizens' lives were in shambles and some even died. The city turned into a honky-tonk town.

Have you ever thought of the impact your life has made during your lifetime?

This will be just minimal compared to all the future effects it will have. Consider this old proverbial rhyme (its origin is probably in the 14th century):

For want of a nail the shoe was lost.
For want of a shoe the horse was lost.
For want of a horse the rider was lost.
For want of a rider the message was lost.
For want of a message the battle was lost.
For want of a battle the kingdom was lost.
And all for the want of a horseshoe nail.

Now reconsider the impact you are having on how you live your life.

We are at the top of the animal world. I believe we are thinkers with the mandate from God to make the right choices in life. He does not interfere with our thinking. Otherwise, we would be mere puppets under his control. If we are not fully responsible for our actions, then we cannot be held accountable for them. It would be close to cheating at solitaire if He didn't let us play our cards out ... alone.

Can God see into the future? From my location, I cannot even start to attempt to answer this question. This is the "radio frequency" limitation we have as humans. Also, we can only think within the boundaries of our ecosystem.

Use all the components that comprise your power to face life's difficulties—you need to address them with your inner strength.

When I was a Marine stationed in Okinawa in 1962, I had to type up a General's handwritten speech. He was a recipient of the Medal of Honor, the highest award given in the military. It requires the approval of the President and Congress.

The speech was about courage. I remember the part that was his personal definition. He said that jumping on a live hand grenade to save your buddies is more of a knee-jerk reaction than courageous. He went on to say that facing the day-to-day difficulties, no matter how hard they might be, was real courage. No argument here. This chapter is called "Your World," but it also includes mine. I can truthfully state that the 50 years I have lived after the five years I spent in the Marine Corps were much more difficult.

Writing this controversial book is also taking courage. In your world of religion, I hope my book gives you comfort and serenity in your beliefs. May it make you a better person and please God that you are on His beautiful Earth. May the goodness you express be spread to others. The Butterfly Effect is a phenomenally powerful tool. Use it well. In my world, I get an out-of-body, tingling feeling of His presence in me.

God bless all those who read and connect with these words. May God bless all humanity. He is within us.

Let us now journey into the world of science and ecology to strengthen our beliefs and the understanding of my theory that there is a oneness of the universe. It has been humanity's incredible minds that have created these breakthroughs.

Science, theology and ecology do co-exist.

CHAPTER 5

THE SCIENCES

Genetics

Let's look at what man has done in the field of genetics. The word "genetics" derives its meaning from the Greek word *genesis*, or "origin."

Have you ever heard of, or pronounced, the word deoxyribonucleic acid?

No wonder, it is merely referred to as DNA! In 1962, the Nobel Prize in Physiology or Medicine was awarded to Francis Crick, James Watson, and Maurice Wilkins. Specifically, the Prize was "for their discoveries concerning the molecular structure of nucleic acids' significance for information transfer in living material." In short, they discovered how the code of DNA inside every cell replicates itself so that all our new cells are exactly the same as the replaced cell, unless there is a rare mutation. It also is the main code inside the sperm and ova that passes on our traits to new offspring.

Dr. Watson wrote a fantastic book on the subject, *The Double Helix* (1968).

The latest estimate of genes in humans is 20,000 to 25,000, and is constantly updated.

Each gene is either dominant or recessive to a specific trait in our bodies (e.g., eye color, hair, etc.). Each parent passes on its gene, either dominant or recessive, and the dominant trait is the determinate for that trait in the offspring. Even your brain is made up of both your parents.

Scientists around the world are now taking and using this discovery to new heights. In 1996, the first animal in the world to be replicated by cloning or DNA manipulation was "Dolly," the sheep. In 2005, "Snoopy the Dog" was cloned at a Korean university.

According to Cliff.com, "More popular among the medical community is the possibility of therapeutic cloning from stem cells promising to bring new hope and stunning potential for a variety of diseases and conditions in the years to come."

Can the cloning of human beings be far away? It's an eerie thought considering possible ramifications. What would God think? Would the cloned individuals have a soul? "The age-old question of 'what came first, the chicken or the egg' was not moot since theoretically, all scientists needed to produce a chicken was one single cell," according to Cliff.com.

Do we have a soul? The first mention of a "soul" appears in Genesis, Chapter Two, Verse Seven: "And the LORD God formed man *of* the dust of the ground, and breathed into his nostrils the breath of life, and the man became a living soul." No duality here; man and soul are one entity. What do you think? Is *thinking* of a new idea and what you *believe* two distinct ways of mental harmony?

Genetics was one of my favorite courses when I attended Brigham Young University. That was nearly 50 years ago, and genetics was at the embryo stage. Yet the course intrigued me almost as much as it did the instructor. He emphatically insisted that we consider the course only the beginning of our knowledge and that we should keep abreast of all new research on the subject. He was right! Genetics has taken off since the 1960s. Below, as listed by Robert T. Gonzalez, are just some of the discoveries and the diseases they could treat:

Epilepsy Gene LG/2:

"Take epilepsy gene LG/2 for example—Epilepsy is

the most common neurological condition in children. The gene discovery was made by a group of researchers at the University of Helsinki led by Dr. Hannes Lohi, who said it will open up many avenues of research that will provide insight into the mechanisms underlying neurological development in the adolescent brain".

Boule, the world's most universal sexy gene:

We say "sexy gene." "By that, we mean a gene specific to sex—the gene Boule is not only responsible for sperm production, it is actually the first known gene to be required for sperm production in species ranging from insects to mammals."

Earlier I wrote that what was most puzzling to me was how two sexes existed without being created by God. Even trees and shrubs, millions of them, are male or female, one or the other. For example, willow trees, aspens, and cottonwoods are either male or female. It is the male plant that produces allogeneic pollen; the females do not produce any pollen.

"MYB-NFIB Fusion gene found in 100% of examined adenoid cystic carcinomas. Fusion genes are created when a chromosome mutation causes two otherwise healthy genes to join together." This is a glandular cancer usually found in the head, neck, and breasts. Yes, gene mutation causes cells to grow into cancer.

These discoveries are just several of many. As you can understand, isolating a gene and its purpose is just the start to cure diseases that are genetically grounded.

As genetics advance, livers, kidneys, and other vital organs will be reproduced in laboratories to replace those that patients urgently need.

About 150 years ago, Charles Darwin (author of *The Origin of the Species,* [1859]), may have unwittingly observed genetics at work. Adaptation and survival of the fittest over countless years permitted him to see it at work on the isolated Galapagos Islands.

It is my belief that the species living today are the recipients of the healthiest and strongest dominate genes passed on from incalculable generations. However, that would only be true if mankind had been isolated from devastating wars and natural disasters.

Gregor Johann Mendel, born in 1822, is considered "The Father of Modern Genetics." He did his research with garden peas. He was able to cross-pollinate different varieties and create new hybrids. Albert Winchester, in his book *Introduction to Genetics*, states that Mendel "... discovered the fundamental basis for the inheritance of characteristics which hold true for all forms of life."

Wikipedia: "Genetic engineering has [made] significant changes in the animal and plant world. It involves the introduction of foreign DNA or synthetic genes into the organism of interest."

The *Drosophila Melanogastic*, or fruit fly, played an important role in the advancement of genetics. The fruit fly has been used extensively, from which a major portion of our knowledge of genetics has been derived. The fly's short life span and large number of offspring permits scientists to rapidly see any genetic changes.

Insulin and human growth hormones are now produced in labs. Humans have altered the genomes of species for thousands of years through artificial selection. The various breeds of dogs are a good example.

"The first field trials of genetically engineering plants occurred in France and the USA in 1986. Tobacco plants were engineered to be resistant to pesticides." (Wikipedia)

Seedless grapes and watermelons are just a few of our modern-day products that have had their genetics modified.

There is hardly a product you buy or eat that has not

been genetically modified (GM).

Thirty years ago GM was a relatively new topic. Who could have imagined how quickly and how so much of our food is now genetically modified?

What amount or percentage of the food you buy and eat is GM? Here is a real shocker: according to Health Freedom Alliance, "It is estimated that about 75 percent of processed foods sold in the U.S. contain at least some genetically modified food ingredients. Unlike many other countries, there is no law in the U.S. requiring the labeling of foods that contain GM ingredients."

Perhaps too often, many years pass before some FDA-approved medications begin to manifest horrible side effects. I believe that cancer cells are created by "outside" genes that enter our bodies and cause a mutation that ignites healthy cells to replicate differently. The malignant cells grow and metastasize to other parts of the body. Now our foods are carrying genetically modified cells that could cause sickness and even cancer.

Here again is the possibility of Newton's law of motion (action-reaction). Better Health Channel describes how genetic modification works: "Genes use chemical messages to instruct the cell to perform its function by making proteins or enzymes. By introducing a foreign gene, scientists prompt the altered cell to make new proteins or enzymes so that the cell performs new functions. For example, the gene that helps a cold water fish survive low temperatures can be inserted into a strawberry to make it frost resistant. The gene may be taken from one animal plant or micro-organism. If the genes are inserted into another species, the resulting organism is referred to as 'transgenetic.'"

Here are some examples that use this genetic technique:

- Crops are genetically engineered to be resistant to insect pests.
- Plants are genetically modified to ensure longer shelf life.
- Farm animals are genetically modified for faster growth rates.

The above are just a few of several that I have researched. Does GM affect the soil or atmosphere? We know that trees make oxygen.

Scientists are tampering with our natural ecology.

TLC (TV Channel; Discovery Communications) lists the 10 products that most use genetic modification:

- Cotton
- Tomatoes
- Papaya
- Rice
- Potatoes
- Corn
- Soy
- Milk
- Canola Oil
- Aspartame (Artificial Sweetener)

Organic foods are ninety-nine percent safe from GM. "Contamination" would only occur if seeds from a nearby GM farm were blown onto organic food farms.

It appears to me that our scientists are creating a whole new "Human Being" with cloned body parts (someday our entire body) and with GM food. We certainly will not be the same as Adam and Eve. And in terms of evolution, it will not be natural ecology that changes us, but a bunch of scientists.

Where will we be in 100 years—if we are still here? We are currently witnessing exponential change. Previously our evolution was a slow process. Is the human body ready for rapid and drastic transformation? While our *bodies* are on an odyssey of their own, how will our *minds* react? In the entire history of mankind, there is no precedent to offer us any inkling of what to expect.

I wonder how God views this progression of changes to His world?

In September 2012, Russia "suspended the import and use of an American GM corn following a [French] study suggesting a link to breast cancer and organ damage. ... Experts at the University of Caen conducted an experiment running for the full lives of rats—two years ... [and] found raised levels of breast cancer, liver and kidney damage." (U.K. Daily Mail, 9/25/12).

There is a compelling argument here for consuming only organic food and promoting GM-free products. But who in America is really concerned? Obesity is now a main cause of adult health problems and our children following the adults' path. Another major reason for our rising cost of health care is our lack of preventive care (diet, exercise, checkups).

"Frankenfoods" may take many decades before they manifest their potentially dangerous side effects. Genetic study of *Drosaphila Melanogastic*, or the fruit fly, may be helpful. Wikipedia states "about 75% of known human disease genes have a recognizable match in the genome of fruit flies, and 50% of fly protein sequences have mammalian homologs."

I do not believe that the ecology of the human body can safely handle rapid changes. *Changes can cause a reaction.* Who wants to roll the dice? But integrated genetic modification goes right down the food chain—from the products that we and the cows, chickens, and pigs eat. GM corn and potatoes presently give us no choice because they are not labeled as GM.

Even viruses are being used. Scientists choose a virus that will invade the target cells but not cause damage or death to the plant or animal. We know how viruses can rapidly mutate and become dangerous. It may take several years for this to occur in what we eat. Personally, I would rather not take the chance.

According to healthfreedoms.org, "Many American organizations have already raised concerns about GM foods, defined as those in which 'foreign' genes—ones not naturally found in the plant—are introduced. The Council for Responsible Genetics, Greenpeace, the Union of Concerned Scientists, the Center for Food Safety, and Consumers Union have all demanded more research and caution. The debate intensified when Britain's Institute of Science in Society launched its own rebellion and convened an independent panel of scientists to rule on the subject. In May 2012, the Institute issued a report: 'The Case for a GM-Free Sustainable World.'"

But the GM deed is done. Good, well-meaning organizations are often "Johnny come Lately's." They react too slowly to affect meaningful change. GM is already deeply embedded in our food chain.

The American people are aware of the new electronic age (iPhones, big screen TVs, etc.), but are largely ignorant about what they are putting into their stomachs.

In 2011, livestrong.com reported GM's negative effects that have already been discovered: "One of the biggest objections to GM foods centers on the unintended potential for harm not only to humans who eat the products, but also other organisms that may consume the crops. Some genetically modified foods, for example, contain genes that increase resistance to certain antibiotics. If this property were transferred to a person eating the food, antibiotics might not have the usual effects against infection."

"Introducing genetic material from one plant to another may result in the introduction of allergenic material from one species into another. Because certain proteins cause more allergic reactions than others, people with severe allergies know what food to avoid. If genes from those foods are introduced into others without appropriate labeling to warn those with allergies, allergic reactions could occur. The introduction of genetic material from Brazil nuts into soybeans was shelved for this reason."

In 2009, the American Academy of Environmental Medicine (AAEM) stated, "several animal studies indicated serious health risks associated with GM foods, including infertility, immune problems, accelerated aging, faulty insulin regulation, and changes in major organs and the gastrointestinal system." The AAEM has asked physicians to advise all patients to avoid GM foods.

With 75 percent of processed foods GM, and with no labeling, how can doctors properly advise patients? It took me four years and several doctors and hospital visits to diagnose my own disease, only to find out it was incurable and they did not know what caused it.

My scenario for the ecology of GM:

Scientists invent genetic engineering to change the DNA of a cell within a plant or animal. We eat the plant or animal without knowing it has been altered. We may not experience an immediate adverse effect, but a catastrophic effect may show up in our offspring sometime in the future. On the other hand, we may have an immediate bad reaction to the food, but we do not know it was caused by the GM food.

There is no end to how far genetically modified foods will go, but we certainly should be well informed and longer test periods should be mandated.

Let us review how any new scientific changes, including GM, come about:

First it starts with a need and an idea that may become a solution to a problem. Then it goes through research that may show promise. It if does not show potential, the project is dropped. If it does, scientists proceed by making a prototype and lab test it. If it passes the lab test, they make more of it and test it on live plants or vegetables, or even humans. If results are positive, it goes through FDA testing stages. If the FDA gives the green light, it goes into manufacturing and sales.

What does this process leave out? Are we going too fast? In the natural world, living things go through a slow evolutionary process and are subjected to *gradual* changes in the environment or ecology. We are not talking about decades, but in many cases centuries or millennia. This natural process and long amount of time are left out of the GM process. This could result in major, unnatural disruptions.

As I mentioned earlier, the atom is basically empty. It gives us an illusion of hardness. Now we have genetically altered animals and plants—and, in the future, perhaps cloned human beings.

We have benefitted from multiple genetic changes, but will we eventually someday suffer unknown or unintended consequences?

Genetics have taken us from new dog breeds, garden peas, "Dolly," replicated organs, and further along the path to probably cloning of humans.

The more we tinker with God's gift of evolution and natural ecology, the more we may be creating havoc with the future of humanity.

Caveat emptor—Buyer beware!

Regeneration and Hybrids

I have sometimes thought how wonderful it would be if we humans could grow a new arm or leg after an accident, illness, or war injury. Several animals have this *regenerative* ability. You might have learned about this in your biology or science class in your high school days.

FactMonster.com lists seven species that have regenerative power:

Lizards who lose all or part of their tails can grow new ones. This is a good escape technique. A lost tail will continue to wiggle, which might distract the predator and give the lizard a chance to escape. Most lizards will have regrown their tail within nine months.

Planarians are flat worms; if cut into pieces, each piece can grow into a new worm.

Sea cucumbers have bodies that can grow to be three feet long. If one is cut into pieces, each piece can become a new sea cucumber.

Sharks continually replace lost teeth. A shark may grow 24,000 teeth in a lifetime.

Spiders can regrow missing legs or parts of legs.

Sponges can be divided. In that case, the cells of the sponge will regrow and combine exactly as before.

Starfish that lose arms can grow new ones; sometimes an entire animal can grow from a single lost arm.

I noted earlier that each of your cells contains the DNA code to replicate your entire body. But unlike the creatures listed above, we humans can only *rejuvenate*—replace old cells with new cells—as demonstrated daily by our skin cells. Your outer skin is replaced about every ten days and the old skin accounts for much of the dust in your home. Old cells are continuously being replaced everywhere in your body. If this rejuvenation process stopped, so would your body.

The discovery of DNA, its mapping, and our increasing knowledge of how DNA works can only be viewed as colossal scientific and medical breakthroughs.

The Club Moss fern has been on Earth for about 300 million years. Club Moss existed before dinosaurs. It is found in shaded areas, especially in Texas and Mexico. It has a special way of rejuvenating. It can hibernate for as long as 50 years. It curls up and winds can carry it vast distances. Once it finds water, it begins growing again. The ecology of the wind transporting it to new and fertile places to grow is critically important for Club Moss's continuous life.

Photosynthesis is the process plants use to convert sunlight into chemical energy. It was previously thought to be something only plants can do. In 2010, a Florida biologist named Sydney Pierce discovered that a species of green sea slug makes its own energy-containing molecules "without having to eat anything." Many scientists now consider the sea slug to be half plant and half animal. It looks like a large leaf with a snail's head.

Humanity, for its own purposes, has been altering plants and animals for thousands of years. Hybrid dogs were bred to achieve characteristics to meet a certain needs, e.g., hunting, killing rodents, guarding, sheepherding, leading the blind and sniffing out drugs and bombs.

More recently, crossbreeding of dogs is done to please owners. Below are several popular dogs and the mating matches:

• Labrador, Retriever, and Poodle = Labradoodle
• Beagle and Poodle = Poogle

- German Shepherd Dog and Poodle = Shepadoodle
- Pomeranian and Yorkshire terrier = Yoranian
- Bulldog and Dalmatian = Bullmation
- Chihuahua and Dachshund = Chihuahuashund

With rare exceptions, most dogs until several hundred years ago were working dogs. Keeping a dog as only a household pet is a more recent phenomenon.

Humans with their co-working dogs constitute an ecological relationship. The phrase "Man's Best Friend" came from a politician-lawyer's closing arguments in a famous trial in Missouri in 1870.

Greek mythology contained many hybrid creatures. Chiron was half man and half horse (a centaur). Perhaps the most famous was the Minotaur, half man and half bull.

Ancient Greece was renown for great philosophers, architecture, and a form of democracy. Their creative spirit was obviously reflected in their mythological gods.

Two additional human-created, weird hybrid creatures deserve brief mention.

First, the *featherless chicken* which eliminates the need for plucking. This creature is a cross between a bare-necked chicken and a regular chicken. Will it need a coat in winter?

Second, the *Zubron*—a cross between a cow and a Polish wisent (European bison). Created in Poland after World War I, the Zubron can withstand extreme cold, is disease resistant, and requires minimal human care.

I also recall my Chinese woman friend who recently visited Tibet. She returned home with a "half-worm, half-grass" hybrid that is used as an ingredient in soup. Its real name is *yartsa gunbu*—a fungus—but colloquially it is called a "worm." The worm is actually a natural phenomenon, and all attempts to farm it have failed. (National Geographic, August 2012)

There seems to be little humans will not do to force natural ecology to fit and satisfy their needs. Sometimes, with sensible and sensitive planning, the results can be beneficial and have minimal impact on nature. Yet it seems that too often humans plunge ahead thoughtlessly, damageing natural ecosystems and contributing to the likelihood that there will be severely detrimental consequences—often unintended—in the future.

Our children are already suffocating under America's massive debt. Where were the "watchdogs" to raise the financial alarm? For many citizens, moral values appear to be declining. Free speech at colleges has almost disappeared under so-called "speech codes." Knowledge itself and many traditional values have become so "fluid" that we need to press the refresh/reset button nearly every day. Or hour. *Click!*

While ancient civilizations and tribes created their various gods, I believe in *one* God—sourced from different religious texts (books). In the final analysis, I think we humans—all of us—just want to have a supreme being.

Geology—Our Home

Geology is the study of the Earth. I will cover its history from its very beginning. The science comprises the study of solid earth, the rocks of which it is comprised, and the process by which it evolves. Geology is extremely helpful to archeologists and paleoscientists in the dating of uncovered findings. It is "the basics" to miners in locating the probable areas in which beneficial ores are found, where it is best to drill for oil, and also in finding precious metals and gems.

I personally consider geology the dessert of courses that I studied in college. In later years as a CPA, I had the thrill of auditing a gold and silver mine in Honduras and an aluminum operation in Messina, New York.

The Earth is about 4.5 billion years old and even at this age it is still cooling off. We see volcanoes spew red-hot lava that has worked its way up through fissures in the Earth's surface from the hot magma below.

The Earth is in a constant state of profound change that goes far beyond what we are familiar with: earthquakes, tsunamis, volcanoes, and weather erosion. You could be sitting on a mountaintop that once was under an ocean or a sea. You could also be swimming in a lake that may have once been part of a mountain range. The Earth is made up of six layers starting with the center core and ending with the outer crust. It is the outer crust that I will briefly and mainly review.

The outer Earth is made up of three rocks: sedimentary, igneous, and metamorphic. Sedimentary rock is created from the sediment of material on the bottom of oceans, seas, and even lakes over millions of years. This would include, but not be limited to, dead sea life and erosion of surrounding land mass. Igneous rock is the crystallization from melt (magma and/or lava). Metamorphic rock is due to heat and pressure that changes the mineral content of the rock and gives it a characteristic fabric.

The outer layer of the Earth is composed of tectonic plates that are constantly moving. The Earth is believed to have been a single land formation surrounded by water long ago. Over millions of years, the single land piece broke up, coming apart, and formed the individual continents. Portions of New Jersey were once part of northwest Africa.

It is the movement of these plates that also created our current mountain ranges as the plates slowly moved and met a land mass that would move the mass upward and cause mountains, such as the Himalayas (one of the younger mountain ranges). Generally, the smallest mountain ranges have been eroded away by weather (Adirondacks).

As you dig or drill into the ground, the denser it becomes. It is presently impossible to dig as far as 50 miles down because the pressure would be too dense.

I mentioned the sedimentary rock formation on the seabed. Our current land mass was once also land that reaped the benefits of long dead vegetation and animal life. Coal came from ancient vegetation crushed by erosion and dense pressure. Oil derives its existence from deceased dinosaurs subjected to the same process. Oil is now being drilled from under the ocean floors that were once landmasses and roamed by these dinosaurs.

As you drive throughout the western states, you can see beautiful stratified mountains of red and grayish colors, which are sedimentary rocks once under water.

While on a geology field trip in the nearby Wasatch Mountains near Provo, Utah, each student in my class was able to find at least one trilobite fossil within about an hour. They are from the arthropods marine life family, and at least 210 million years old.

There is no doubt that it took billions of years for the Earth to reach this beautiful and bountiful condition. It took billions of years to cool down and also to create the perfect atmosphere for we humans to breathe. It evolved and was not created in a day or two. It may well have been the way God wanted it to be: eventually livable for us.

There are many ways scientists have to date the approximate time of their discoveries. Here are a select few:

Carbon-14 (C-14) Dating:

It is used on organic matter, whereby the living species takes in carbon 14 during its lifetime. Upon death, the dead organic matter emits C-14 at a constant rate.

Knowing the amount of C-14 at death and the amount remaining when found can, with a little math, determine the organic matter's age.

Stratigraphy:

If the fossil is found in a predated sedimentary layer of stone, the age is usually quite clear. This can also be used in the reverse way.

Radioactive Material:

This is similar in method as C-14 dating, except the item must have some kind of radioactive content in it. An organization, World's Earth Science, describes the technology: "[the] technique takes advantage of radioactive decay, whereby a radioactive form of an element is converted into another radioactive isotope or non-radioactive product at a regular rate. Math does the rest."

There are dozens of methods used by scientists to date and cross-date their discoveries and they share the results with colleagues around the world. What does the Chicxulub crater and dinosaurs have in common?

The crater was created by a huge asteroid that killed the dinosaurs 65 million years ago. It caused dust to circle the Earth, cut off sunlight, which eventually killed the vegetation and therefore with no food, no life.

From Wikipedia: "In March 2010, following extensive analysis of the available evidence covering 20 years' worth of data spanning the fields of paleontology, geochemistry, climate modeling, geophysics, and sedimentology, 41 international experts from 33 institutions reviewed available evidence and concluded that the impact at Chicxulub triggered the mass extinctions at the K-Pg boundary including those of dinosaurs."

The area they are referring to is on the Yucatan Peninsula and off the coast of Mexico. Glen Penfield, a geophysicist looking for petroleum during the late 1970s, initially discovered it.

Throughout recent history, it has been the work of geologists and associated scientists who have been the principle discoverers of essential minerals and precious items including coal, copper, zinc, gold, diamonds, oil, natural gas, etc.

Once again, I believe in the "untampered" evolutionary process and that God is part of it and not the cause of it (Butterfly Effect). How He is part of it is theorized later.

Next we will cover what it is on Earth that makes life possible.

Earth's Atmosphere

The atmosphere is perhaps the most important element for plant and animal life on Earth. Like other fundamental components I will discuss (ocean, sun, moon, gravity), the atmosphere evolved naturally and not for the sole purpose of suiting humanity. We evolved into the way we are because of general, natural evolution that required a livable atmosphere.

The atmosphere provides several benefits beyond enabling us to live. It protects our planet from being hit by debris from outer space, such as asteroids, meteorites, and meteors. These objects are usually burned up by friction when they begin penetrating the atmosphere. Without this protection, Earth would have pockmarks and craters similar to our moon's surface.

The only reason that the sunlight goes around corners is because we have an atmosphere. The moon is all dark or all light because it does not have an atmosphere.

Leaving heavy salt behind, ocean water evaporates upward and forms clouds that produce rain and snow. Wind created by our atmosphere spreads the liquid around the Earth.

The air and its elements enable birds, insects, and planes to fly.

The ozone layer protects the Earth from the sun's harmful ultraviolet rays. The sun's rays create rainbows, lovely sunrises and sunsets, blue skies, and cloud formations.

The atmosphere controls Earth's temperatures and the winds that spread seeds, move sailboats, and break up stagnate air.

What is the atmosphere composed of and how far up into space does it go?

The constituents of the atmosphere are nitrogen, 78.09 percent; oxygen, 20.95 percent; argon, .93 percent, and several small amounts of a variety of other gasses. The oxygen component is mainly produced by blue-green algae.

There are several layers within Earth's atmosphere before we reach the vacuum of space around 43 miles above Earth.

The closest layer to the Earth is the troposphere, reaching up to a height of seven miles. Following in order are the stratosphere to 30 miles up; the mesosphere to 50 miles up; the thermosphere to 440 miles up, and finally the exosphere that leads to space at about 440 miles from Earth's surface.

There are certain layers such as the ionosphere within the main regions that exhibit characteristic properties. "Auroras, or northern and southern lights, appear in the thermosphere. The ionosphere is in the range (50 to 400 miles) that contains a huge concentration of electrically charged particles (ions). These particles are responsible for reflecting radio signals important to telecommunications." (From infoplease.com)

Sciencemadesimple.com tells us why the sky is blue: It

is "due to Rayleigh scattering. As light moves through the atmosphere, most of the longer wavelengths pass straight through. Little of the red, orange, and yellow light is affected by the air. However, much of the shorter wavelength light is absorbed by the gas molecules. The absorbed blue light is then radiated in different directions. It gets scattered all around the sky. Whichever direction you look, some of the scattered blue light reaches you. Since you see the blue light from everywhere overhead, the sky looks blue."

When I was a little boy I was told that the ocean's reflection made the sky blue. It was probably best that way—I would not have understood the Rayleigh scattering.

As I got older, I pondered. How did the sun and atmosphere make a rainbow, the most splendid masterpiece of nature?

A short summary of the physics behind the rainbow:

The basic colors and their wavelengths are important to know. The longest wavelength is red (630 – 700 nm), followed in descending order by orange (590 -630 nm), yellow (560 – 590 nm), green (590 - 560), blue (450 - 490 nm), indigo (420 - 440 nm), and violet (400 - 450 nm). When the sunlight rays go through suspended water droplets in the atmosphere, which serves as a reflector, the raindrops then act as prisms. When the wavelengths exit the droplets, they cause a dispersion of light that reflects in their distinct colors.

I often see rainbows displayed in my dining room. When sunlight hits the crystal glasses in my dining room cabinet, the prisms of a rainbow are reflected on the crystal.

How was the atmosphere formed?

According to science-at-home.org, "The first atmosphere came from volcanoes and was mostly water and carbon dioxide. When it cooled down it rained and made the oceans and a lot of the carbon dioxide dissolved. Later some type of algae (blue-green) started making oxygen until eventually the atmosphere was like it is today."

Oceans

The universe began about 14 billion years ago. The Earth formed about 4.5 billion years ago and the oceans formed about 4 billion years ago. Water came from the atmosphere and created the oceans. Water laden with dissolved minerals also entered the oceans as runoff from the land. The minerals contained the salt that makes seawater salty. Oceans comprise seventy per cent of the earth's surface. For a billion years, the oceans remained fairly constant.

"Ocean ecology" is often described as the study of the effects that humans and technology have on the oceans. This popular definition emphasizes the role of humans in harming the environment. My definition of "ocean ecology" also includes all organisms in the ocean and the ocean's effect on Earth's weather and on all of Earth's living things.

Scientists have only begun to uncover the myriad secrets of the oceans' depths.

Life on Earth likely began in a hydrothermal (warm) spring and on a deep ocean floor where it was protected from ultraviolet radiation. It was probably a location that could not dry out, was not too cold or too hot, and was neither too acidic nor too alkaline. "Life began there, inside the membranous froth of minerals that surrounded these hydrothermal springs, and the minerals themselves catalyzed the first chemical reactions that made proteins and nucleic acids—precursors to the earliest living things. Every kind of life, including us, carries the mark of that first catalysis in the form of tiny mineral clusters at the centers of our enzymes, still doing the jobs they began so long ago." (From americanscientist.com).

The first living organisms were one-cell prokaryotes that lived about 3.5 billion years ago. They are similar to present-day bacteria in the depths of the ocean.

Fish evolved as the first of six animal world classes. Reptiles, amphibians, birds, mammals (including humans), and invertebrates followed. I will focus on animal life in or involved with the oceans.

Water is a necessity for life on Earth. NASA's *Curiosity* and other probes search for present or signs of past water on Mars as a clue to possible life forms.

How many oceans are there on Earth? While named oceans are all connected and there is truly only one World Ocean, we still refer to four: Pacific, Atlantic, Indian, and Arctic.

Fish came before dinosaurs.

Below I briefly describe ten fascinating prehistoric fish that are still alive today. When an organism or group of organisms disappear from the fossil record for one or more periods and then are found again, they are called Lazarus taxa (plural) or a Lazarus taxon (singular).

Coelacanth:

The 400-million year old Coelacanth was believed to have become extinct along with the dinosaurs, but in 1938 a live specimen was caught in South Africa. Other Coela-canths were caught afterward and photographed. Over six feet in length, they taste terrible and therefore are rarely caught. They are a critically endangered species.

Dinosaur Eel (Polypterus Senegalus):

Also known as Dragon Fish, these 14-inch African fish are obviously named for their highly reptilian appearance. They first appeared 65 to 145 million years ago. Their spiked backs resemble some dinosaurs' backs. They belong to the bichir fish family—not to the eel family. Often sold as exotic pets, Dragon Fish will eat all the smaller fish in the same tank, and are highly prone to escaping.

Alligator Gar:

The largest freshwater fish in North America, the Gator Gar can reach 13 feet in length and weigh over 350 pounds. It can breathe air and survive out of water for two hours. It is carnivorous and ambushes smaller prey in the streams, rivers, and bayous of the southeastern United States. Its long jaws with a double row of sharp teeth obviously resemble the alligator. There have been no documented attacks on humans. The Gar has existed for 100 million years and is presently overfished in many areas.

Sawfish:

Also known as the Carpenter Fish, the Sawfish can grow to a length of 23 feet and weigh 1,000 pounds. All Sawfish are considered critically endangered. The saw-like rostrum is used as a tool for digging up crustaceans on the ocean floor. Sawfish are 100-million year survivors, but are disappearing because of industrial fishing. BBC's "Planet Dinosaur" tells us that dinosaurs' "fossilized teeth are commonly found with the remains of the giant sawfish. In 2005, a spinosaur fossil was found with a sawfish vertebra stuck in a tooth socket and another discovered in 2008 had a fragment of a sawfish barb apparently embedded in its jaw."

Arapaima:

The freshwater Amazonian arapaima can also be traced back to the age of dinosaurs. Mature fish are 6 to 14 feet long and weigh up to 440 pounds. The arapaima can breathe air. It eats other fish as well as birds near the surface of water. It is also another endangered "living fossil."

Sturgeon:

There are 26 species in the sturgeon *Acipenseridae* family. The Beluga of the Caspian Sea is the largest freshwater fish on our planet—weighing up to 3,500 pounds and 24 feet in length. Beluga caviar is considered the tastiest roe (fish eggs) in the world. Beluga is approaching extinction just in the last 15 years because of heavy fishing.

Frilled Shark:

Chlamydoselachus anguineus has an eel-like body and is about 6 feet in length. It can swallow whole prey over one-half its own size. Some scientists date this deep-sea fish found in the Atlantic and Pacific back 250 million years, to the Mesozoic period. It is considered "Near Threatened."

Arowana:

The freshwater Arowana dates back 150 million years, to the Jurassic period. They can be found in Africa, the Amazon, Australia, and Asia. They are carnivorous surface feeders, capable of leaping over 6 feet into the air to catch birds. They are obligatory air breathers. They are about a yard long and weigh up to 38 pounds. Arowana are considered "lucky" because they resemble the auspicious Chinese Dragon. This fish is considered an endangered species.

Lancetfish:

With watery and gelatinous flesh, the lancetfish or "Wolf of the Sea" is not a commercial target. They are over six feet long and can weigh 10 pounds. It has few enemies, has lancet-like teeth, and is known as a cannibal. Found in all oceans, it dates back to 40 million years.

Hagfish:

Hag eels (not really eels) are the only animal known to have a skull but not a vertebral column. The hagfish dates from 50 millions years ago. It emits a slime that clogs the gills of potential predator fish, so it has few enemies. The hagfish's eyes can detect light but cannot see images. Hag eels are 18 inches to 4 feet in length and are basically scavengers. Some scientists believe this fish has two brains.

* * * *

The highest mountain in the world is under the ocean. While Mount Everest is over 29,000 feet high, there is a mountain under the Pacific Ocean that is nearly 33,000 feet tall. The deepest part of the ocean is about 36,000 feet (nearly seven miles).

With oceans covering over two-thirds of the Earth's surface, and at one point deeper than our highest mountain, the oceans indeed offer a vast area for ecological study.

The oceans contain about 20,000 fish species and nearly two million animal and plant species including small life forms like worms and jellyfish.

There are three mammals that live in the ocean: whales, dolphins, and porpoises.

Writing about this variety of sea life reminds me of a personal "fish story"—the time I caught an octopus in the East China Sea back in 1962, over 50 years ago. I had no idea of how to get the three-foot, eight-armed blob off my hook. An Okinawan on board did it for me and he was elated when I told him he could have the creature. The octopus is a delicacy to Okinawans and many Europeans on the Mediterranean coast.

The world derives many economic benefits from its oceans. We fish, swim, dive, cruise and tour. Fishing provides income. One-sixth of American jobs are ocean-related.

The oceans provide us with snow and rainfall required for our survival.

Cargo boats around the world ship countless products. Without this shipping we would still be in the Dark Ages—if we existed at all. Actually this is a moot point since the first living organism came from the bottom of the ocean.

Weather

The combination of Earth, atmosphere and oceans create our weather.

God gets blamed for weather disasters more often than anyone or anything else. Even the insurance companies refer to destructive weather damage as "Acts of God." The science of weather is called meteorology. Our atmosphere is composed of 78% nitrogen, 21% oxygen, and 1% argon. Our survival depends on just the right mixture of these atoms.

There are myriad sayings about the weather. The Weather Channel (on television, the Internet, and mobile devices) talks about the weather 24/7, but no one can do anything about it.

I have an example of the weather versus a tree in my backyard. When I bought my present house, I knew the lot had been neglected and needed fresh landscaping. As a retiree, I had the time to tackle most of this job. I hired professional help when needed. I faced a variety of scattered trees and bushes, plentiful weeds, a man-made fishpond without fish, and a grassy area that needed replacing.

Escrow closed in the late Fall, so I did my planning during the winter months.

A lonely, bare twelve-foot high tree stood in the

middle of the ragged grass area. I could not identify it because it had no leaves. In the spring I replaced the sod in the grass area, working around the tree. A female friend suggested that I remove the tree. I disagreed. When the leaves grew out, I saw I had a beautiful pink Mimosa.

After several years, the Mimosa grew a large branch on its leeward side, opposite to the direction of the wind. The tree was counter-balancing itself to maintain a vertical trunk. It is now thirty feet high and the centerpiece of my backyard.

As I look through a large glass sliding door, I see mountains in the background while in the foreground my yard teems with plant and animal life. Birds soaring in mid-August, a favorite month. The elevation is over 5,000 feet. My house is adjacent to Prescott National Forest.

Ecology is not a snapshot, but a motion picture—it is always changing. Weather is also not a snapshot and is the cause of most of the changes in our biological world. Noah's flood, ice ages, hurricanes, lightening-caused forest fires, drought, and you-name-it are all weather related.

National Geographic's September 2012 cover story was "Texas: the New Dust Bowl."

The Earth and its weather have their own ecosystem. Earlier I said our solar system (Earth, Moon, and Sun) is similarly related. At some point in the Earth's 4.5 billion years, there was just one land mass surrounded by water.

As tectonic plates moved, separate land continents were created. As they continued to shift, deserts became jungles and vice-versa. The Sahara desert still has one of the world's largest aquifers under it.

During the Ice Age, glaciers moved south; when they melted, the fresh water filled five enormous holes now called the Great Lakes. Huge rocks were deposited at the place of melting. Geologists can easily see these rocks are not congruent with the surrounding area. When you drive along the highways and see a large stone that looks out of place, it could be one left over from the Ice Age.

As mountains rose from movement of tectonic plates, snow fell on them. When the snow melted, rivers were created.

Our recent weather changes are small potatoes compared to our past. Just look what happened to dinosaurs 65 million years ago. Also, consider the huge amount of deep-water oil wells and reserves. Wells are drilled thousands of feet below sea level. The ocean's floors were once dry land with plant and animal life.

Poor farmers in northern Russia are getting rich as melting arctic ice reveals huge, once-buried mammoths. The farmers pull out the ivory tusks and sell them. Imagine how many events had to occur over millions of years to fill the farmers' wallets.

Weather has the distinction of being both fast and slow. Mountain erosion is slow and a lightning strike is, well, *as fast as lightning.*

Humans cannot replicate the beauty of the exposed, sedimentary rock formations of the mountains, mesas and canyons of many Western states. Solid earth and rocks was the raw product; the weather carved out its features like a sculpture. Mountains, made from sedimentary material over millions of years, were once under water. Sedona, Arizona has as especially splendid display of God's work. Within a day's drive, I can view Monument Valley, the Grand Canyon, Zion National Park, and Utah's Bryce Canyon. The sight of evergreen trees growing on the sides of the red rock strata is a magical contrast.

I was born during a thunderstorm. Is that the reason I love watching storms and cloud formations? My sister Carol and her husband Gus came from New Jersey to visit me at my new home. It was their first trip to Arizona. As we sat on my patio deck, I watched their awestruck faces as they watched the sun go down. I have to assume that New Jersey sunsets do not compare.

The haboob sand and dust storms of the Phoenix, Arizona area are a sharp contrast to the northern part of the State.

The ever-present nature of weather, both good and bad, perhaps has the most diverse and widespread effect on all living things than any other ecosystem.

Daily weather is reported on TV and radio. The biggest computers are dedicated to help in this process. However, even with all this man and machine analysis, Mother Nature still fools us. Countless times in Prescott, when the weather forecast was for "sunny and partly cloudy," I watched rain pouring down on my neighborhood.

Actually, keeping the big picture in mind, we all could be dead in one second. No need for a long Armageddon. God could become upset with how unappreciative we are of His beautiful world and end it all. Along with the weather, nothing is truly predictable. It is no wonder that Eastern religions teach us to rid ourselves of wants (they depend on the future), and that Eckhart Tolle tells us to think in the "now."

A side note: It would be difficult, if not presumptuous, to arrange this book chronologically. The ideas, events, systems and processes I discuss all are happening at the same time and are all interrelated. We certainly do not have a "mind" like God, so why do we attempt to solve the impossible? Because that is our human nature. Accomplishment is accomplishment. Inventiveness, problem solving, discovery and enlightenment give us all a great feeling when we succeed.

How does weather affect our minds? Most of us become a little depressed when it is gloomy and raining outside—unless we are in a drought and our crops need saving. How about the rainbow, one of the many gifts of Mother Nature?

On the other hand, most of us have experienced devastating weather at its worst—floods, lightening-related deaths, snowstorms, hurricanes, tornados, etc. I have personally made a few hundred-mile detours because of rain and snow.

While at college in Provo, Utah, my wife and I were invited to a classmate's home in Rockland, Idaho for Thanksgiving. His father came down to pick us up. As we drove north, we hit a snowstorm that forced us to make a 150-mile detour around the mountains. My wife was pregnant with our first child, and my classmate's father didn't want to take any chances.

When we were living in Las Vegas, we had a boat docked on Lake Mead. On two separate occasions we encountered severe weather. The first time, we drove an hour and half to our boat. After we got the tarp off the boat we saw dangerous-looking black clouds over the lake. We quickly covered the boat back up and headed home over flooded roads. A police roadblock eventually stopped us. We turned around and headed north towards Utah, then cut across the Valley of Fire and back to our house— at least a one hundred mile detour.

The second time was also on Lake Mead. One magnificent morning, my sister Carol and her husband Gus set off with my wife Gerry and me for a cruise across the Lake and up the Colorado River towards the Grand Canyon. In the early afternoon, we saw ominous clouds gathering overhead. At this point, we were about 20 miles up river. I decided to head back.

Storms form quickly in Nevada, and often without warning. When we re-entered Lake Mead, wind and waves pushed straight into us. Our forward progress was cut by more than half and we were heaved by dual currents. Waves were coming overhead and it was raining hard. Our remaining fuel was burning up rapidly because of the boat's fight for forward movement. The gas gauge needle bounced up and down because of the turbulence. As best as I could read it, the tank was somewhere between 1/8 and empty.

I could see land in the distance, but Echo Bay was far to the north. We struggled northwest in the increasing darkness. I was mindful I could not steer my boat, *Tooner Schooner,* without power. Everyone was tense. I was eager to reach the marina as soon as possible, but I had to hold the throttle at a slow speed to conserve gas.

Finally, we reached Echo Bay. I eased the boat into the slip and we all simultaneously yelled a big cheer. There was only one man to greet us at the drenched marina. He shouted, "What the hell are you folks doing out there? No one else is on the Lake!"

In my log, which I still have, I wrote: "December 4, 1998—Carol, Gus, Gerry and I—eight hours bad weather to Colorado." To say the least, my entry was exceptionally concise.

My weather experiences are tame compared to those occurring daily around the world. However, the world would not be as it is today without the exact weather of the past. In fact, the evolution of mankind would have been much different, if it had occurred at all. That would make us the sole beneficiary of God.

Sciences of Our Unusual World

From the Earth to the outer space, there exist highly unusual phenomena that are truly curious and amazing.

My personal belief about the appearance of things is that nothing is funny looking, weird, scary, or eerie in the universe or on Earth. The various forms and looks of most organisms are, of course, the result of evolution. How something appears to its predators often determines its survival. Our human features are designed for functional purposes—ears, eyes, nose, teeth, legs, arms, etc. The sun, our atmosphere, and gravitational pull all influence our outward appearance. We exercise, diet, dress, and use cosmetics to enhance our looks, but I am sure we would still look strange to aliens as we would to them.

"Many animals transform almost beyond recognition in the course of their lives," says newscientist.com (January 2012). "Caterpillars become butterflies and tadpoles become frogs, and if we couldn't watch them do so we might not even suspect that the two stages were the same creature." Darwin's *Origin of the Species* is about adaptation and survival of the species which provides the empirical evidence of what we are now witnessing.

New biological species are always being discovered. The evolutionary process never rests.

A newly found single-celled organism, *Mesodinum chamaelon*, is a unique mixture of animal and plant. Newscientist.com: "They engulf other microorganisms, generally algae.... The two then form a partnership. The algae produce sugars by photosynthesis, while the *Mesodinium* protects them and carries them around.... The division between plants and animals is collapsing completely."

The three human races are becoming increasingly

64

intermarried, producing mixed offspring. Will we ultimately develop into a world of one race and one belief in the same God, and eventually become more tolerant?

To further the proposition that the animal world evolved in appearance in order to survive, I have selected several examples:

The Giant Coconut Crab:

The world's largest land-based arthropod (invertebrate animal) lives in coastal areas and forest regions of the Indo-Pacific islands. It is most active at night, can weigh up to nine pounds, and can reach nearly three feet in length (including leg span).

Aye-aye:

The world's largest nocturnal primate, the aye-aye lemur (*Daubentonia madagascariensis*), is only found on Madagascar. Aye-aye has an "unusual method of finding food; it taps on trees to find grubs, then gnaws holes in the wood using its forward slanting incisors to create a small hole in which it inserts its narrow middle finger to pull the grubs out.... From an ecological point of view the aye-aye fills the niche of a woodpecker, as it is capable of penetrating wood to extract the invertebrates within." (Wikipedia) An Aye-aye can weigh up to four pounds, and can be two feet in length including their tail.

Giraffe Weevil:

This newly found species (*Trachelophorus giraffa*), also of Madagascar, is named for its giraffe-like neck. The male's body is about an inch long. The long neck is an evolutionary adaptation and is used for fighting and nest building.

Stick Bug (or Stick Insect):

Phasmatodea, also known as walking sticks or stick bugs, are very difficult to spot because of their adaptation of natural camouflage. They look remarkably like a very small, broken off piece of twig. I have seen these bugs outside my home in Prescott, Arizona.

Glass Frog:

While their topside color is green, the glass frog's underside is almost transparent. The internal organs can be seen through the skin.

Mata Mata:

This freshwater turtle of South America has a large flattened head with many flaps of skin. *Chelus fimbriata* reach a length of about 20 inches, and weigh as much as 30 pounds.

With the Earth having a wealth of all manner of nutrients, it is no surprise we have such a diverse multitude of plant and animal life. The nutrients provide for the adaptation process that Darwin advanced.

Black Holes

Moving away from Earth's "unusual" critters, I am compelled to mention the Black Holes in the universe.

I have been in the business world all my life, from my college degrees through my professional career, but my true love was the sciences. I never took a course in physics, but have been reading about it for over 50 years. It is not that it comes easy to me, but it is fascinating.

To more fully explain Black Hole intricacies, I will quote (below) from Wikipedia. There have been sci-fi movies based on them, but far fetched.

A black hole is a region of space-time where gravity prevents anything, including light, from escaping. The theory of general relativity predicts that a sufficiently compact mass will deform space-time to form a black hole. Around a black hole there is a mathematically defined surface called an event horizon that marks the point of no return. It is called "black" because it absorbs all the light that hits the horizon, reflecting nothing, just like a perfect black body in thermodynamics. Quantum mechanics predicts that black holes emit radiation like a black body with a finite temperature. This temperature is inversely proportional to the mass of the black hole, making it difficult to observe this radiation for black holes of stellar mass or greater.

More simply put, the Black Hole is not empty. On the contrary, it is an extreme mass that is so dense that not even the light can escape its gravitational force. It was created by the death of a huge dying star or supernova. The supernova exploded and the remnants collapsed, becoming a black hole. We cannot observe them even with our strongest telescopes, but know of their existence by their gravitational pull on passing objects.

This description is also very abbreviated, but sufficient for our purposes. God did not just plant Black Holes around the universe; the laws of physics created them. They are also observed in our own galaxy. If we cannot really "see" them, how do we know they exist? "Despite its invisible interior, the presence of a Black Hole can be inferred through its interaction with other matter and with light and other electromagnetic radiation." In short, if there are other stars in the area, the Black Hole's orbit will be affected and its location determined.

Does anything I have written so far trigger any new ideas? I will be glad to take a royalty on any new profitable invention that you discover. Many years ago, after watching sci-fi movies and the development of rocket propulsion, I came up with a new idea regarding space travel:

If a spaceship were equipped with a powerful device that aimed its positive gravitational pull towards the destination planet, and its negative force towards the Earth, the force of gravity would do its thing. The ship's speed would be dramatic, and no fuel would be required. However, after learning more about gravitation, I have abandoned this idea. No royalties here.

The seed for this idea may have occurred 50 years ago when I was studying geology. I learned that the North and South Poles switch places every half-million years or so. While digging up sedimentary layers of rock, geologists noticed that magnetization of metallic content was in a different direction from higher layers when compared to older and deeper layers. The last Pole reversal was 780,000 years ago. Are we overdue?

* * * *

Have you heard about animals that reproduce asexually? No fun here (at least not that we know of). Amoebas, which you probably learned about in high school, are single-cell organisms that reproduce by dividing, thereby producing two smaller copies of itself. Greenflies can clone themselves once every 20 minutes. When I lived in Las Vegas, there was a small lake with a sign that read "Warning: Amoebas Present—No Swimming." These organisms can be dangerous to your health.

Then there is the black widow spider that eats its male after mating. No fun here either. That is another problem I have with human sex and evolution—why the enjoyment of reproduction? Perhaps God's hand was in the evolutionary process?

* * * *

I read a book long ago about a Northwest African tribe that traded gold for salt. The trading activities occurred over a century ago. The salt was as precious to the tribe as the gold was to the Muslims of Mecca, Saudi

Arabia. Neither group could understand each other's language, and the tribe feared direct contact. The trading would proceed as follows:

The Arabs would place a certain amount of block salt along a riverbank and then retreat out of sight. The tribesmen would approach the salt and leave the amount of gold they thought the salt was worth, and retreat. The Arabs would approach the gold; if they thought it was not enough gold, they would take back some of the salt. The tribesmen would then return to the riverbank, put down some more gold, and once again retreat. This process continued until each party was satisfied. This trading continued until the Arabs had all the gold they wanted.

Since gold is so precious, why can't we just chemically manufacture it? There are many answers to this question. I believe that "chemistry deals with the interactions of the outer electrons of the atom, and to make gold you would have to alter the nuclei of the atoms.... Rearranging the components of the nucleus would not be impossible, even if we do not have the technology to do that now, but it would take an enormous amount of energy, so we would end up paying much more for gold than it costs to mine from the Earth." (From answers.yahoo.com)

If we could manufacture gold at a lower cost than mining it, how would the world economies be affected? Supply and demand determines price, so the effect would be highly disruptive. As you can see, it is not as easy as mixing ingredients in the kitchen or alchemy.

Back in 1971, when President Nixon severed the link between the U.S. dollar and gold, all foreign currencies rose as the U.S. dollar got hammered and our purchasing power was sharply reduced abroad. I was a CPA on Wall Street at the time and one of my clients had operations throughout the world. I had to help them re-translate their financial statement. This was a huge task for me since their account-ants did not know how to do it.

Many economists, 40 years later, believe our economy has suffered ever since the U.S. dollar became a fiat currency.

If gold became cheaper to manufacture instead of mining it, it would have a ripple effect throughout the world. Gold prices would plummet just on the announcement of the new process and gold investors would lose on their holdings.

Shareholders in gold mining companies would suffer huge losses and mining employees would be out of work. The ecology of the economic world would be in turmoil. Tampering with ecosystems can be perilous. Remember Sir Isaac Newton's law of motion: for every *action* this is a corresponding opposite equal *reaction.*

The sentiment in financial markets would drive econ-omies to depression levels, and debt-laden governments would be incapable of rescuing their respective economies. We would have a worldwide financial disaster on our hands.

It is the supply side of gold, not the demand side, which is more sensitive.

If gold were lying around the ground in our own backyards, it would be as commonplace as dirt, but more attractive.

* * * *

In the science of astronomy, the aurora borealis, or Northern Lights, is as beautiful as it is an astronomical rarity. Wikipedia tells us that the aurora "is a natural light display in the sky particularly in the high latitude (Arctic and Antarctic) regions caused by the collision of energetic-charged particles with atoms in the high altitude atmosphere.... The Cree called them the 'Dance of the Spirits.'"

I have only seen pictures of them and they are awesome.

* * *

Earlier in this chapter I wrote about evolution of some of the strangest looking creatures on Earth. Evolution was not concerned with our ego and personal desire to be attractive. Evolution focused on our survival. But we human beings do love mirrors and "looking good." Many male animals (lions, bulls) fight for the right to mate and the winner gets the reward. It is the best (strongest) of their species that win—not the merely attractive—and the winners carry forth the best DNA.

When I was doing business in Columbia, South America, I attended a dinner at a wealthy family's house. During the course of the evening, I had a conversation with one of the locals. He told me about visiting a tribe in the jungles of Columbia. He saw a monkey in a giant boiling pot, bouncing up and down and hairless as a baby, as it cooked over an open fire. The monkey was the tribe's dinner and the local was a guest.

He said he always brought small gifts that were most pleasing to the tribe. I asked what the gifts might be. He replied, "They loved mirrors and cups." After a little thought, I could see the relevance—these items are our own daily needs.

I have personally interacted with many people around the world. Despite our different cultures, religions, races, and ethnicities, I found that all of us have much in common.

* * * *

As you read about ecology in the next chapter, keep in mind the sciences we have covered. It will serve you well to recognize the oneness of it all.

CHAPTER 6

ECOLOGY

After years of research and intensive thinking, I found that ecology *in its fullest expression* offers the best and most trusted path to a valid and complete Theory of Everything (TOE).

A truly comprehensive understanding of ecology—a *robust ecology* fully and imaginatively examined and understood—reveals *the oneness of ecology itself.*

Ecology manifests oneness, and oneness manifests ecology.

Oneness expresses itself best and completely through an all-embracing, absolutely inclusive ecology that excludes nothing in the universe ... *and is the universe.*

When my "robust ecology" is fully grasped and implemented, I almost see it as *res ipsa loquitur*—the thing speaking for itself—and that "thing" is everything. That thing, a oneness, when finally understood through my expanded use of a fully dynamic ecology, can only be a singular, totally inter-connected oneness.

Ecology reflects oneness, oneness reflects ecology—because they are the same: One.

Einstein discovered his theory of special and general relativity through advanced mathematics and physics, neither of which is required to understand my theory.

About fifty years ago, when I was in college, I took a course in zoology. The concept of "ecology" was tucked away in the last chapter of the course textbook. The realm of ecology goes far beyond the conventional biological world on Earth.

In my book, ecology reaches the ends of the universe. Included in this chapter are the physical masses in the universe and how they affect each other as well as the biological world.

I also include the "spirit" of God in everything which makes it so wonderful. Because God is part of—and integral to—my scientific theory of ecology, I assert that theological beliefs and the physical world are united. The Butterfly Effect on future events cannot be left out. All areas are inter-connected, which I call its oneness.

The pages that follow, as do the above ideas, cover a vast array of ecology—both good and some not so good.

On some topics of ecology, scientific subjects will be presented to clarify relationships and validation.

Try to keep an open and hungry mind. Most of all do your own thinking as you read.

The Big Picture—The Universe

Earlier, I wrote briefly about the Big Bang theory that is presently accepted as the beginning of our known universe. The theory states that time, space, and matter were created from extremely small and dense pieces of "matter" that exploded.

I have added to this theory. I posit that God (the Almighty or whatever greater power you believe in) was also part of the Big Bang and therefore our ecology.

Where else could He have been if no space or matter existed before the explosion?

I also wrote that once the expansion stops (it is right now expanding at an accelerated speed), the laws of gravity will pull everything back to its original state. This is called the Big Crunch theory in physical cosmology.

To justify to theologians that God and the universe are eternal, the Big Bang will once again occur at this point. This process will never end.

How does ecology play its part in this scenario?

On its greatest scale, the Earth was created within the laws of physics. But this was just the beginning of our livable Earth, as I explained in previous chapters.

Continue to look for the oneness of ecology as we look at the universe.

Ever-present Ecology

Webster's New Universal Unabridged Dictionary defines ecology as "the branch of biology dealing with the relationships and interactions between organisms and their environment, including other organisms." From Knowledge Project of the Nature Education: "Ecology is inextricably intertwined with the evolution history of organisms." Genetics passes information from one generation to the next, information that enhances biology's ability to access this record to better understand the origins of species.

The mammals, and subsequently *Homo sapiens,* were survivors of the asteroid. Many species of mammals lived and stored food in burrows and did not have the need to consume that which dinosaurs required. This became the age of the mammal.

Earlier I wrote about the ecology of our sun, moon, and the Earth. Even the Big Bang (Singularity Theory—mass, space and time created at once), and due to their relativity (mass, space, and time), could be considered ecological.

When she was a little girl, I once told my daughter, Tammy, that perhaps an atom inside of her could be the center of the universe. At the time the size of the universe was unknown and they were theorizing on the innermost mass of an atom.

In algebra you are taught that everything can be cut in

half. They even used the analogy of the race between the tortoise and the hare. During the race when the hare was far ahead, the question was, "If the tortoise cuts the distance between the two in half, when will the tortoise catch up to the hare?" Answer: never. The tortoise will always, by definition, be halfway behind. In order to arrive at true and corresponding answers in life, one must have a full understanding of the questions.

A college roommate, Cortland Walker, told me the hare and tortoise story. I had to take algebra in my college freshman year without the benefit of any previous high school course on the subject.

I was having a difficult time because some of the homework problems did not make sense to me. Cortland was a math major and by way of his example, I was able to grasp the equation like the tortoise and the hare.

Three years ago, I began thinking about the ecology of everything—as Webster's dictionary defines Ecology—"dealing with the relationships and interactions between organisms and their environment, including other organisms." I define the word "environment" as the universe.

"In ecology, predation is a mechanism of population control. Thus, when the number of predators is scarce, the number of preys should rise. When this happens, the predators would be able to reproduce more and possibly change their hunting habits. As the number of predators rise, the number of preys decline. This results in food scarcity for predators that can eventually lead to the death of many predators." (From Biology-Online.org) The predation mechanism also demonstrates the Butterfly Effect.

A good example of a down to earth ecological relationship is the balance of wolves and deer out West. If mankind were to kill off too many wolves, the deer population would elevate, and deer would die due to too little foliage to eat in the wolf-deer domain. In many parts of our country, laws have been passed to preserve endangered species. We humans should be put on the endangered species list for all the wars we have experienced.

How does God feel when He sees all the killing--wars, murders, genocides, etc.?

If I were God I would be very angry and disappointed. How many wars could be justified in light of the Ten Commandments? How can a President in many speeches say, "God Bless America?" How can we justify this hypocrisy? Historians Will and Ariel Durant wrote in their book, *Lessons of History,* that in the past 2,000 years, there was some kind of war was going on in 90 percent of those 2,000 years. In *The Worldly Philosophers* by Robert L. Heilbroner, the author covers the Malthusian doctrine:
" ... population, unless unchecked, grows at a greater rate than the means of subsistence [and] would result in [world] starvation." Thomas Malthus's theory was not acceptable by his peers then and has not been proven since.

It is clear to this writer that the ecology of the animal kingdom is in sync except for humankind. What do you think?

You have heard of the phrase, "Thinking outside the box." It is quite a hackneyed expression now. I believe in pure thinking—no boxes.

Thinking outside the box doesn't tell you who is looking from inside the box. They could be biased on many topics that would taint their opinions. I prefer the word "Think." Be your own person.

I recall taking a religious course at BYU given by a professor of philosophy.

I remember asking the professor how I could believe in any religion if it required me to step inside that "box" with its dogma, tenets, rituals, history and books, etc. to be able to accept that religion. I further stated that I would need to step inside their "toolbox" to be a believer, but this would prejudice my thinking. I cannot remember his answer, probably because I was not convinced.

While watching a science fiction movie, do you have a hard time distinguishing between science and fiction? Even the science may be a dichotomy—fact verses theory. The movie itself is mostly an illusion since the atoms are almost empty. When an elated person says that they feel like they are walking on air, they are closer to the truth than not.

Most of the damage to the Earth's ecology was in full swing before the average person knew or heard of the word ecology. We had put toxins into the air (acid rain, ozone layer) and dumped toxic materials onto the Earth. If we really needed oil, our country would enter a period of economic mayhem. We would drill, drill, and drill!

Like Ben Franklin wrote in *Poor Richard's Almanac,* "Hunger never saw bad bread."

Even though Germany had a non-aggression pact with Russia during the Second World War, they attacked Russia's Western front to get to the oil. The Russians believed the pact, and that is the main reason the Russians suffered 20 million casualties—they were never prepared for war.

Humans killing humans (90% wartime in the last 2,000 years) is the predatory ecology that puts us high on the chart of this category, even with our "superior" brain.

To expand on the ecology of the wolf-deer-food: when one species disappears, its predators can no longer eat it, and the species that disappeared will no longer be dining on the prey it used to eat. Changes in these populations affect others (the Butterfly Effect). This is a kill-eat relationship (oppositional relativity). There are also symbiotic relationships, like the honeybee and the flower.

The bee gets nectar to make honey from the flower and contributes back to the flower by spreading the pollen so that the flower can reproduce.

Another positive ecological relationship is called commercialism. "For example, a small fish called a Pilot fish follows underneath a shark and when the shark eats something the Pilot fish eats the scrap pieces of the shark's original kill." (From "Blue Planet," a BBC Documentary, 2001.)

The epitome of positive ecological relationships is mating and a need for two sexes. The Bible gives the answer—God's creation. But the author and dating, or age, of Genesis is in question or unknown. According to "The True Origin Archive" and an article in *Evolutionary Theories on Gender and Sexual Reproduction* by Brad Harrut, Ph.D. and Bert Thompson, Ph.D., "Evolutionists freely admit that the origin of the sexual process remains one of the most difficult problems in Biology."

This has been the most baffling question to me and I don't know that I will get the answer in my lifetime. The answer may be similar to how Rene Descartes reached the Philosophical answer to the question, "How do I know I exist?" His answer was simple: "I think, therefore I am." How do these mental gymnastics answer the illusion of seeing objects around us that we are almost empty?

Do God and mankind have an ecological relationship? I have no idea; so let's go to the ecology of my backyard.

I have a single-family home with one-half acre of land in Prescott, Arizona. It is nestled in the hills and small mountains, and not densely populated. There is such a diverse selection of animals that roam through, or live in, my backyard that I enjoy it like a geographic magazine.

I have roadrunners, garden snakes, bull snakes, small lizards, rodents, tarantulas, skunks, rabbits, scorpions, javelina, coyotes, deer, Mexican grey wolves, mountain lions (one sighting), raccoons, brown recluse spiders, birds, and an endless list of insects. Having been bitten once by a brown recluse spider while sleeping, I know from personal experience that it hurts for several months. As you can see from my vast, diverse list, some are the predators and some are the prey (e.g., bull snakes and rodents).

I watch the roadrunner and garden snake in action from my large sliding glass window. The roadrunner comes out from out of nowhere and pounces on the unsuspecting 18-inch snake in a grassy area. He picks it up and drags it even closer to me, onto the floor of my cement patio. There he slashes at it repeatedly on the hard concrete. After the snake is mostly dead, the triumphant bird carries it off to a secluded place to eat his dinner-prey.

Another ecological pair at work is the bull snake and the rodents. I watched the bull snake crawl under the opening under my tool shed in search of the rodents that make the shed their home. The rodents (ground squirrels, prairie dogs, etc.) are my "enemies" because they love to eat my plants and flowers.

Last year, I saw four or five skunks nearly every night (I have motion-sensing lights). They would sniff around for insects and even tarantulas. This year, I haven't seen either. Ecology at work?

When I bought my house in central Arizona I planted grass in the backyard. It didn't take long for the rabbits to feed on the fresh new grass. At first I was angry to have these invaders attack my grass. It didn't take long for me to accept them as part of my active yard.

Now I look at the rabbits playing on the lawn as they frolic during the mating season. I look at the beautiful morning sunrise over the mountains. I feel a vibrant sensation go through my entire body as I feel God in me (omnipresent). Collectively, they give me the desire to write this book. If just one reader of this book has a positive reaction, than I feel complete.

I had an incident last year with two tarantulas in the darkness of the night. As I walked out my sliding glass door, I felt something fall on my head and strands of a web hit my face. I instantly brushed the hairy mass off my head and face and stepped back into the house. I turned on the patio lights on and saw two tarantulas, one on the ground, and one still in the web.

The symbiotic relationship of the ecosystem that benefits both parties is baffling. A good example is the relationship of the bee-nectar flower honey exchange. Without the pollen, the flower couldn't reproduce. This, like two sexes, is beyond my ability to credit evolution, but swings towards creation. Both are good examples of "what came first, the chicken, or the egg?"

Ecology and Entropy

"Entropy" is not a word you hear very often. When I wrote that change causes change, I was using one of the ways to describe entropy. The concept becomes a little more complicated because entropy emerged via philosophers and mathematicians. John Nash (see *A Beautiful Mind*, 2001) used it in his work on Game Theory. I studied Game Theory in graduate school (Operations Research) and found it very challenging.

Wikipedia: "Entropy is a property of thermodynamic systems. A thermodynamic system is any physical object or region of space that can be described by its thermodynamic qualities such as temperature, pressure, volume and density." This supports my position on the ecology of the universe.

Wikipedia continues: "In a thermodynamic system pressure differences, density differences, and temperature differences all tend to equalize over time." For example, imagine a room. The room contains a drinking glass and there is melting ice in the glass. Think of the room, glass, and melting ice as *one system*. "The difference in temperature between the warm room and the cold glass of water and ice is equalized as heat from the room is transferred to the cooler ice and water mixture. Over time the temperature of the glass and its contents and the temperature of the room achieve balance." A naked man at the North Pole would soon equal the temperature surrounding him and turn to a frozen (and dead) human being. Deciduous trees lose their leaves in the winter months.

Entropy covers the law of order and disorder. A cold glass of water in a warm room is a state of *disorder.*

As the glass of cold water equals the temperature of the room, it becomes *orderly*. It is headed for oneness.

Here lies the relationship between ecology and entropy. They both seek a balance—a natural balance. Order.

Entropy also shows us the arrow of the passing of time. When we see things around us change, we know it is the passing of time and in the direction of the future.

Horizontal Ecology

Thus far I have been reviewing ecology that moves with time—cause, then effect; or if this happens, then that happens, and so on. But we also found a rational solution to "what came first, the chicken or the egg." In the evolutionary process, the chicken evolved and then came the egg. All creatures and plants are in the big envelope of evolution. *We all evolved.* If we accept the Garden of Eden creation in which a male human came first, then a woman, then the offspring—no "babies first" here.

Do not bother to look up "horizontal ecology" elsewhere in this book or anywhere. It only appears in this chapter, and my meaning is this: countless changes are happening all at the same time that in effect create a new world instantaneously, and lead to a future we cannot see until it happens.

Several examples of simultaneous change:

A man goes for a walk on a path in the woods. He comes to a fork or "Y" in the path. Will he take the path on the left, or the one on the right? He chooses the one on the right. Two miles away, at the same moment, a brown bear chooses between two paths and selects the one on his left. Bear and man meet on the same path and the bear mauls the man to death. Unintentionally, the bear made the right choice and had a feast. Unknowing, the man made the wrong choice.

On a sunny afternoon, a generous and wealthy woman decides to take a walk around a nearby park. Later that afternoon she has scheduled an appointment with her attorney to rewrite her will. She plans to leave all her fortune to charity and one homeless woman she has befriended. While on her walk she is struck by a bolt of lightening which instantly ends her life. Her large fortune ends up in the wrong hands, going to her undeserving and selfish relatives.

A drastic "extinction event" could affect the whole human race in just one second. God could kill all of us—without reference to the Book of Revelation. He could become totally exasperated by our genocides, greed and the damage we are doing to the world's ecology. People do not work or function by the laws of science (Newton's laws, Einstein's general relativity, etc.), but by the choices they make.

As the saying goes, "You can't walk across the same river twice." You will never live this one second—right now—of your life again.

Life is not a snapshot, but a motion picture. As you go through your ever- changing world, you adjust your choices to fit the circumstances you face, and therefore you are part of the ever-changing world. Change evokes change.

Within your own body, you do not breathe in the same air every time you take a breath, nor does the same blood run through your own heart with every heartbeat. And as the clock ticks, you are constantly getting older. But the clock is just a man-made, arbitrary way of informing us of the time.

When you couple the changes triggered by horizontal ecology (events that change at the same time) with those changing vertically ("this-then-that" as time passes), the future becomes a totally new world. Change initiates change. For example, if you lost your job today, you would counteract the loss with another change.

This is another way of thinking about the Butterfly Effect.

None of this is profound, it is just reality. It does stress the importance of making well thought out decisions that ultimately will have a beneficial impact on your future and the future of others. The cliché, "Think before you leap," comes to mind.

Most of us tend to blame our past for our current status or state of mind, but we must be realistic regarding how we got here. Could you be the one mostly to blame for making wrong choices? We cannot do anything about the past or the future. But we can do something about the "now"—the present.

It is a highly uncertain world we live in, and we are

always on an uncertain voyage. But the choices we make today (now) are certain.

Think long and well and make your life enjoyable and productive.

Oddities of the Human Body's Ecosystem

As previously stated, I accept evolution as God's way of introducing mankind on Earth. Since the creation of Adam and Eve in the Bible (around 5,000 to 6,000 years ago), we have found and developed almost endless amounts of scientific knowledge that leans toward long-term evolution. I have already briefly touched on some of these science-based findings.

The ecosystem of the body gives us more information regarding our ecology and the world. The sciences in the previous chapter highlight how important everything must be on the Earth to enable our survival.

According to Erin Allday of the *San Francisco Chronicle,* "The human body carries more than 100 trillion bacteria—up to five pounds of ... tiny single-celled organisms."

Rediscovering Biology (learner.org) reports that "the bacteria we possess are an ecological community; thus, the principles of community ecology and evolution are vital to understanding how these bacteria (both the benign and the potentially harmful) live within us. Each bacterium species is adapted to the habitat and ecological niche it fills, existing in somewhat of an ecological balance. The balance helps thwart the invasion of pathogens which must compete with resident bacteria for nutrients and space. Resident bacteria also produce antimicrobial proteins called *bacteriocins* which inhibit the growth of related species."

Many of the bacteria are critical for our survival. This suggests we evolved with an ecosystem that was necessary for us to survive.

Why do we have toenails and appendices? Why do males have teats and even get breast cancer? Oddities without answers.

If we were created in a "Garden of Eden," we must have begun the evolutionary process immediately to get that bacterium. We also know we must have adjusted to all the ecological changes to get us here today.

Through the centuries our bodies adapted and evolved, eventually developing a built-in immunity to several diseases. When European explorers first landed in the New World, they introduced diseases that were new to the natives. In the Hawaiian Islands natives sickened and died from small pox, an old disease that was "new" for the victims.

Adam and Eve would have had vulnerable immune systems unless they evolved. Disease dates back to the dawn of humankind, but yet the human body survived. We learned how to hunt, plant, create language, build houses and significantly more. Medicines and medical treatments evolved from herbs, plants, and shamans until reaching our present day health care system.

Health and Ecology

What would the world population be today without centuries of progress in medical treatment? Humanity's condition might be unrecognizable.

We have evolved from witch doctors and ancient traditional "cures" to today's wonderful diagnostic equipment, highly educated physicians, and a comprehensive understanding of the human body's complexities. We live longer and enjoy better care along the way.

Modern medical scientists scour the jungles in search

of plants and animal life offering possible new cures for disease. Even our ordinary aspirin's history goes back thousands of years. Hippocrates (460-377 BC) used powder made from the bark of the salicylate-rich Willow Tree for pain relief treatments.

In the early 1800's, scientists determined that salicin indeed gave relief. After combining salicin with other chemical ingredients, Bayer branded the product as "Aspirin" and began selling it around the world by 1900.

Caution! Do not eat bark from your backyard willow. Consult with your doctor. And do not worry—drug companies are probably not hurting the environment by chopping down willows.

Who would have thought that snake venom could save your life? Besthealthsecret.com tells us "A purified form of Malaysian Pit Viper venom, registered with the trademark 'Arvin,' has been used successfully to thin blood. Not only does it dissolve blood clots, it also triggers the human body's own clot dissolving mechanism." All those little atoms forming DNA ultimately had a previously hidden use in medicine.

Now for a short, ecological, butterfly effect story in medical action:

A little boy or girl has a mother who dies from a rare disease. When the child grows up, he or she vows to cure the disease by becoming a medical research scientist. After years of schooling and lab work, he or she discovers a cure for the rare disease, which eventually saves many lives.

I have already written about cloning of human organs from stem cells and related ongoing research. The United States is far behind other countries because of our government's inadequate funding of research.

By now you know I am writing about the *ecology within each science*, but more importantly, how the ecology of science affects the world around it and is part of the ecology of the universe.

For example, if Einstein had died from a disease that could have been cured by a doctor, we would not be the beneficiaries of Einstein's discoveries about the universe and ultimately the Big Bang Theory.

This scenario shows us that the state of Einstein's health allowed him to advance scientific knowledge.

Going one more step, let us say that Einstein was saved by some medical discovery like using the venom of a snake. We now have an ecological connection between a poisonous snake and theoretical physics, albeit far-fetched.

The human body is, indeed, a masterpiece. Consider how often a 25-year old house needs to be repaired, including the appliances. Under normal circumstances, I am sure an average house requires more maintenance than an average 25-year old human body.

How about your car? The age of the average car on the road is about 11 years old. Some children have not yet reached puberty at age 11. A 25-year old car is ready for the junkyard, while we have not reached our peak.

Most inventions and manufactured goods are designed with planned obsolescence, while humans possess a built-in immune system. No doubt about it, we are one magnificent creation.

Earlier I hypothesized about how disappointed God must be with all the wars, murders and atrocities humans commit, but we also are not taking good care of our bodies. According to the Centers for Disease Control and Prevention (CDC), nearly 36 percent—over one-third—of U.S. adults are obese. The American Academy of Child and Adolescent Psychiatry reports that up to 33 percent of children and adolescents are, well, fat. Annual medical costs associated with all obesity are estimated at $147 billion and rising—a profound ecological effect on our country's health and economy.

Many of America's current trends are reminiscent of the fall of the Roman Empire—an empire that died from within, making it vulnerable to the outside enemies. Our government's huge debt, the erosion of our cultural values (TV, movies, entertainment), a dominant myopic preference (live now, pay later) and the faltering health of our citizens point to a rather bleak future. I presently see no change in action to stop this gloomy outcome from happening.

Do you think God wants to interfere with the mess we are in? If He intervened, it would be His solution, not ours, and there would be no human accountability or justice.

The consensus of anthropologists is that we were mostly savages before we became hunters and gatherers. Eventually, civilization came to some groups.

Over the eons, our bodies adapted to climate changes by clothing, better diets and migration.

If the body adapts to the ecology around it, then so does the total makeup of its components. My theory on cancers: When the body is subjected to unnatural changes (e.g., substance abuse, medications, repeated sunburns, GM foods) it wants to refuse to accommodate these changes. The DNA of the cells gets the message from the changing environment. Over a period of time, the DNA code reacts by replicating new cells differently—as cancerous cells. DNA is merely performing its ecological mission: adaptation.

The sheer mass of evidence from mankind's ancient remains around the world makes it difficult for me to believe in one Garden of Eden within a time frame of 5,000 to 6,000 years ago. How can intelligent people recon-

cile the assertions of the Bible and overwhelming archeological and anthropological data?

Whatever the truth is, we should be pleased with our
beautiful bodies.

Ecology's Present-day Problems

While ecology studies have significantly progressed in
the last five decades, we mostly read or hear about *negative*
ecological issues: destruction of coral reefs, acid rain, the
disappearing ozone layer, nuclear waste disposal, and all
manner of pollution. Fortunately, most human-made catastrophes have prompted positive remedial action.

In my own lifetime, we have seen the creation of a
"philosophy of ecology."

According to ecologyandsociety.org, "A good philosophical understanding of ecology is important for a
number of reasons. First, ecology is an important and
fascinating branch of biology, with distinctive philosophical
issues. Second, ecology is only one small step away from
urgent political, ethical, and management decisions about
how best to live in an apparently fragile and increasingly
degraded environment. Third, properly conceived, philosophy of ecology can contribute directly to our understanding of ecology and to its advancement."

Personally, I don't believe Ecology and Society's rea-
soning is sufficiently persuasive. It reminds me of a U.S.
president announcing a new Czar for drugs or some other
current hot issue. And second, I believe ecology goes far
beyond the branches of biology—indeed, ecology includes
the total universe.

I think it would be more effective to educate the
public about *all* the concerns of worldwide ecology. It is
urgent that our government stops underperforming while

Madison Avenue roars along in high gear

You have likely seen television commercials about prescription drugs. Scary warnings regarding possible side effects take more time to disclose than the description of a drug's benefits. Warnings are delivered by pleasant, saccharine voices. I saw a magazine ad for a new anti-coagulant that had one page for the ad—followed by three pages of dire warnings.

After I had been taking a prescription for three years, another side effect was discovered. A new warning was quickly added to the label on the drug bottle.

Our human body is the first "wonder of the world" and operates as an awesome ecological machine. It rejuvenates, has its own built in immune system, gives us signs when something is wrong, and operates as a single entity. When we become ill, the body-machine is so complex that we often must visit several physician specialists to receive a final diagnosis and treatment.

The human brain—so far—is better than any computer, but in too many ways we under-use it. Increasingly we are using computers or artificial intelligence (AI) to "think" for us. I sometimes wonder if artificial intelligence is an oxymoron—like "giant shrimp."

Many children today are not learning multiplication tables (e.g., 12 x 12 = 144) or cursive writing. I am well aware we have entered a new age of technology, but basic math and writing skills are still important. I would like to see our public schools add a comprehensive ecology class for the lower grades. Too many schools are dropping gym or physical education classes. Children are becoming fatter and less fit. What else will be eliminated? A study should be made between our educational system and the students as an ecological system.

The ongoing problem of the disposal of nuclear waste

is not going away soon. Have you heard of the financial disaster at Yucca Mountain in Nevada that cost taxpayers approximately $15 billion and then was aborted?

According to AP (January 1, 2012), "Many Americans became newly aware of the presence of tens of thousands of tons of spent fuel at more than 70 nuclear power plant sites around this country—and of the fact that the United States currently has no physical capacity to do anything with this spent fuel other than to continue to leave it at the sites where it was first generated."

The Yucca Mountain project was developed to be a nuclear waste repository, but the Obama administration decided to halt work there in 2009.

It all had to do with geology.

If the planners had hired geologists to thoroughly study the geologic history and formation before they began the project, they would never have chosen the site.

"Tuff" is a type of rock consisting of volcanic ash. According to Wikipedia, "The formation that makes up Yucca Mountain was created by several large eruptions from a caldera volcano and is composed of alternating layers of ignimbrite (welded tuff), non- welded tuff, and semi-welded tuff. The volcanic units have been tilted along fault lines, thus forming the current ridgeline called Yucca Mountain. In addition to these faults, Yucca Mountain is crisscrossed by fractures, many of which formed when the volcano units cooled."

In other words, the Yucca Mountain area is capable of experiencing an earthquake or volcanic eruption at any time. Nuclear waste would be spewed into our environment. Jet streams in the upper atmosphere generally blow from west to east.

Goodbye, $15 billion ($15,000,000,000.00).

Another ongoing project since 1980 is the cleaning up

all the toxic waste that companies dumped onto our land for decades. Toxic waste can cause cancer. Movies have been made about waste materials seeping into water and making people very sick.

"Superfund" is the name for the Comprehensive Environmental Response, Compensation, and Liability Act of 1980 (CERCLA). It was designed to clean up sites contaminated with hazardous substances. Wikipedia: "As of November 29, 2010, there were 1,280 superfund sites on the [U.S.] National Priorities List. Six-two new sites have been proposed." Every state in the U.S. is on the list. The current or future damage will never be fully known.

From my own personal experience I know that the administrative process is awfully long and the cleanup is even longer. In Vermont, I was president of a manufacturing company in that was included on the list. We never dumped any of our waste, but engaged a company that others were using to dispose of spent oil used in our manufacturing process. It was rather small amount and all manifested with our contractor. Meetings with lawyers went on for eighteen months and were still going on when I retired. No substantive action was taken during that time.

In most cases of human-caused pollution, the harmful causes were discovered long after the events took place.

What is happening today that could harm us tomorrow? No one knows.

Pulitzer Prize-winning Propublica.org (Sept. 28, 2012) describes another toxic waste that has proliferated over the last 60 years. "The waste—the byproduct of oil and gas drilling—was described in regulatory documents as a benign mixture of salt and water. Yet the dangers of injection are well known: In accidents dating back to the 1960s, toxic materials have bubbled up to the surface or escaped, contaminating aquifers that store supplies of drinking water. There are now more than 150,000 Class 2 wells in 33 states, into which oil and gas drillers have injected at least 10 trillion gallons of fluid."

The fine website of the World Wildlife Fund (WFF) (wwf.panda.org) lists the five worst environmental blunders of all time:

Wasting water
Overfishing
Toxics and pollution
Invasive species
Global warming

Under invasive species, WWF tells us that

As a race we're pretty good at this and nowhere is it better proven than us taking species from one place on the planet and putting them in another. Fantastic. Admittedly, the first few times this happened we didn't know better, but then (and you will be shocked to discover) we kept on doing it.

The usual scenario for introducing a species goes something like this (this is a simplified explanation)... We come along to some corner of the world, usually an island, and see something we don't like. But we do know that in another part of the world there's something that eats this thing we don't like in this new place. So we bring it over. And – if we're lucky – it may eat the thing we don't like. But it also eats the things we do like. And a lot more besides. The problem being that this new bad-thing-eater no longer has any predators in the new place to keep it in check. So it runs rampant. It goes, if you'll pardon the pun, wild.

Below are specific examples of mankind's disruption of areas of the Earth:

In 1954, the Nile Perch was introduced to Africa's Lake Victoria to compensate for a huge decline in native fish caused by over-fishing. Instead of helping, the Perch caused the extinction of more than 200 native fish species.

The Small Indian Mongoose was brought to many Pacific islands (e.g., Hawaii) to control rodents. However, the new invader wiped out numerous native birds, reptiles and amphibians.

In 1980, I purchased a home in Franklin Lakes, New Jersey. The Gypsy Moth had been "introduced" to the northeastern U.S. in 1869 by a professor trying to interbreed silkworms with moths. Some of the moths escaped. Over a hundred years later, Gypsy Moths caused me to lose perfectly healthy oak trees around my house. Because I was busy commuting to Dallas, I was largely unaware of the dreadful damage the pests were doing to my trees—until it was too late.

Accidental introductions have occurred throughout the world for centuries. The Black Plague of Europe was one of the most deadly diseases in history. Carried by fleas on black rats from Asia to Italy, the Black Death spread rapidly throughout Europe. In the mid-14th century, 25 million people perished in a five-year period.

Yet some transplants have proven beneficial. Originating in China, peaches are now abundant in many parts of the world. Tomatoes came from the Andes. Pumpkins, corn, and tobacco were introduced to the Old World from the Americas.

Is God part of, or silent to, these ecological changes? My theory, as covered earlier, is that God does not tinker with anything on Earth.

Wasting water is another major ecological blunder by mankind. One-third of the world's population is living in nations that have water shortages.

"In fact," says WWF, "we're so bad at managing our freshwater resources that in the process we have created deserts, poisoned millions of hectares of land with salt and killed entire lakes—in some cases we are even making them disappear."

When I lived in Palm Springs, I occasionally noticed the odor of dead fish from the Salton Sea, about 50 miles away. The Salton Sea is the biggest lake in California (by area, not depth), but the runoff of chemicals in fertilizers from farming around the lake has largely destroyed this once beautiful resort area.

WWF continues, "Mismanagement of freshwater resources is causing floods where there were never any floods, droughts where there were never any droughts."

First envisioned in 1919, the controversial Three Gorges Dam in Mainland China was finally completed in 2012. It was designed to stop flooding that had been going on for centuries. It's the world's largest power station in terms of installed capacity.

Personally, I wish they had finished the dam before 1985. I made about a dozen business trips to Nanjing, Anyang, and Beijing over a 10-year period. In 1985, during one of many meetings, we had to use candlelight because of electricity rationing. I also slept in unheated rooms and endured only cold water to wash in the winter months.

The Three Gorges Dam caused 1.3 million citizens along the Yangtze River to be displaced. WWF adds that the dam "flooded archeological and cultural sites as well as causing significant ecological changes, including an increased risk of landslides." Archeologists tried to save artifacts before the flooding occurred.

I think enough is known about the air pollution we have created from factories and vehicles, as well as the toxins we have dumped into Earth's water and land. The creation of the Super Fund (Comprehensive Environmental Response, Compensation, and Liability Act of 1980 [CERCLA]) to clean up toxic waste dumps and require lower pollution emissions from vehicles is a big step in right direction, but not all nations are cooperating.

How many species are being lost each year? No one really knows. The loss of just one butterfly can theoretically cause drastic changes in the world over millions of years. What will be the impact of all species that have been lost and are dying out today? Collectively, every living and declining species—including humanity—has an incalculable influence on the future. Therefore, no one can predict the future. There are alarming statements by WWF (panda.org) about the loss of species:

Biologists estimate there are between 5 and 15 million species of plants, animals, and micro-organisms on Earth today, of which only about 1.5 million have been described and named. The estimated total includes around 300,000 plant species, between 4 and 8 million insects, and about 50,000 vertebrate species (of which about 10,000 birds and 4,000 are mammals). .... Today, about 23% of mammals and 12% of birds are considered as threatened....

Why are species disappearing? Global biodiversity is being lost much faster than natural extinction due to changes in land use, unsustainable use of natural resources, invasive alien species, climate change and pollution among others. .... Species loss is also compounded by the ongoing growth of human population and unsustainable consumer lifestyles; ... waste and pollutants; urban development [and] international conflict.

Least talked about and known are the ecological problems caused by overfishing. Some fishing is downright foolish. Shark fin soup is a delicacy in China and other areas of the world. For many years, fishing vessels would catch live sharks, cut off their fins, and then toss the sharks "back into the ocean, unable to swim, hunt or survive." (Wikipedia.org) This is now illegal, but the soup can be found. It is the fishing, not the eating, that is illegal.

For the Chinese, many traditional foods are believed

to have health values. Bird's nest soup has been served in China for over 400 years. The primary ingredients are saliva nests built by cave Swiftlets (birds). The gelatinous soup is expensive, ranging from $30 to $100 per bowl. It supposedly aids digestion, increases libido, and alleviates asthma. "A kilogram of white nest can cost up to $2,000 USD, and a kilogram of red nests can cost up to $10,000 USD." (Wikipedia.org)

There is hardly any creature safe from sacrifice for traditional food in China. On my last trip there, I ate Yak and a worm-vegetable soup. I could not say what it really was.

Now back to the overfishing problems. WWF Global, an environmental organization primarily formed for saving the Earth's ecology, tells us "the global fishing fleet is 2.5 times larger than what the oceans can sustainably support. Already 52% of the world's fisheries are fully exploited and 24% are over-exploited, depleted or recovering from collapse. .... And the coup-de-grace of it all is that we are getting less food from the sea. We're landing smaller, younger fish; we're wiping out entire fish populations." (wwf.panda.org)

My first realization of how harmful humanity has been to the atmosphere occurred when I visited Los Angeles in 1958 for two months. Heavy smog caused my throat and eyes to burn. I noticed that most cars had their windows rolled up. In 1954, L.A. schools were closed for some days because of thick smog, and children were kept indoors.

Many changes have been made since the 1950s to alleviate the smog predicament (unleaded gas, catalytic converters, etc.). People residing in "The City of Angels" were one step away from wearing gas masks. For me, smog had destroyed my image of "Sunny California."

What Southern Californians were doing to pollute the

air is now being done on a worldwide scale. The major present-day problems we have created are ozone depletion and the greenhouse effect.

From the 1930s to 1980s we were using chloro-fluorocarbons (CFC's) in everything from air conditioners to aerosol sprays. In the 1980s scientists determined that CFC was depleting the Earth's ozone layers, and CFC's were banned internationally.

Professor Bruce E. Johansen asserts, "Because CFC's remain in the stratosphere for up to 100 years, they will deplete ozone long after industrial production of the chemicals ceases." The maximum loss will occur between 2010 and 2019.

The greenhouse gasses in the Earth's atmosphere collect heat and light from the sun. They are natural and needed to control our temperature, but too many gasses would cause the Earth to get too warm.

From thinkquest.org:

Global warming is affecting many parts of the world. Global warming makes the sea rise, and when the sea rises, the water covers many low land islands. This is a big problem for many of the plants, animals, and people on islands. The water covers the plants and causes some of them to die. When they die, the animals lose a source of food, along with their habitat. Although animals have a better ability to adapt to what happens than plants do, they may die also. When the plants and animals die, people lose two sources of food, plant food and animal food. They may also lose their homes. As a result, they would also have to leave the area or die. This would be called a break in the food chain, or a chain reaction, one thing happening that leads to another and so on.

Ozone loss also causes acid rain that destroys almost everything it touches.

The Goddard Institution for space studies identified the science behind the widening of the ozone holes in the

Arctic and Antarctica.

Dr. Johansen further writes, "The [Goddard] team found that the greenhouse effect was responsible not only for heating the lower atmosphere, but also for cooling the upper atmosphere. The cooling poses problems for ozone molecules, which are most unstable at low temperatures. Based on the team's model, the buildup of greenhouse gases could chill the high atmosphere near the poles by as much as 8 degrees C. to 10 degrees C.

Richard A. Kerr in 1998 describes some of the researchers' conclusion: "Unprecedented stratospheric cold is driving the extreme ozone destruction. Some of the high altitude chill may be a counterintuitive effect of the accompanying greenhouse gases that seem to be warming the lower atmosphere. The colder the stratosphere, the greater the destruction of ozone by CFC's."

This unexpected consequence reminds me of genetically modified foods. After eating GM food for years, will we later discover there are dire consequences? Will we ever learn from (1) the basic law of physics—Sir Isaac Newton's "laws of motion," (2) Darwin's adaptation, and (3) plain old "cause and effect?"

Or that (4) *every change causes change?* Darwin's discovery of adaptation in the Galapagos Islands showed us that change could take a long time, while the other three "laws" can make things happen immediately and have an ever-lasting multiplier effect.

The U.S. government cannot be the watchdog over everything and it is already too large. American society has acquired a state of mind that demands instant gratification and worries about consequences later. Big businesses are giving the people what they want and Madison Avenue sells the idea that discretionary needs are really necessities.

Let's return to the bigger picture—what this book is

all about. I am seeking to understand (1) the universe from its beginning (The Big Bang) as a part of ecology, (2) the sciences that are relevant to ecology, and (3) God (omnipresent, omnipotent, and omniscient)—the universe's creator.

I call this study macro-ecology—the ecologies of all the above. The term macro-ecology appears in wikipedia, but my use of the term greatly expands the term's scope.

Since we are humans, I am writing from a human perspective. We humans are not at the top; we are of the animal world, but only part of it. Since we are the only living things with a genuine capacity to harm the planet, we comprise the gravest hazard to "spaceship Earth."

With the changes we have made in the last century—changes that affect everything simultaneously *and* in the future—our planet Earth must be given full consideration.

Extinct and Endangered Organisms

Even without man's intervention, the ecology of Earth has had its own disasters. However, these events also abide by the ecology of the laws of physics.

Five major mass extinctions have occurred during the Earth's 4.5 billion year existence. The most recent one, about 65 million years ago, wiped out the dinosaurs. Scientists surmise that as much as 99.9% of all once-living species are now extinct. Floods, the Earth's cooling, volcanoes, and perhaps the impact of a huge asteroid caused the older major extinctions. The extinction of species has always been part of the evolutionary process. All the major extinctions were caused by natural events in an ecological system.

Geologists have numerous tools to determine prehistoric dates, including cross-referencing with other fields of science (archeology, paleontology, et al.). Geology was my favorite class in college and I continued to follow fresh developments in the science.

This is similar to the stock market's ecology. You could have all blue chip, triple A-rated stocks in your portfolio, but still suffer great losses if there is a catastrophic event like the Great Depression of the 1930's. All animals can be hit by immense forces out of their control—including average human investors who will suffer from stock market crashes.

Were any of these major extinction events related to the great flood and Noah's Ark in the Bible? I would say it is highly unlikely. There was massive, worldwide flooding when the ocean's level rose because of melting glaciers—but that flooding occurred over 400 million years ago.

Below are several animals that have recently become extinct:

Tasmanian Tiger

The last Tasmanian tiger in existence died in a zoo on September 7, 1936. Hunting and an expanding human population contributed to its demise. It was the largest marsupial (pouched) carnivore of modern times. Both males and females had a pouch. The tiger was about six feet in length nose to tail, and weighed approximately 65 pounds.

Golden Toad

The last sighting was in May 1989 in Costa Rica. Some believe the toads perished because of a chytrid fungal epidemic, habitat loss and/or climate change.

Caribbean Monk Seal

This seal officially disappeared from Earth in 2008. It was last seen in 1952. Human activity was the likely cause of its extinction. Over-fishing depleted the seals' food supply, and the seals themselves were over-hunted for their oil.

Pyrenean Ibex

The last Pyrenean Ibex died in the year 2000. It was one of four Ibex subspecies. Competition with other wild and domestic ungulates (hoofed mammals) accelerated its demise. The *UK Telegraph* (Jan. 31, 2009) reported that scientists took skin samples from the last living Ibex and cloned it, creating a female. The clone died shortly after its birth because of lung defects (a common occurrence in the cloning process). It was the first time an *extinct* animal had been cloned. This amazing, breakthrough technique may be the only way we can preserve species nearing extinction.

Javan Tiger

Last seen in 1972, this tiger could reach eight feet in length and weigh up to 310 pounds. Burgeoning human population extirpated the tiger through "agriculture encroachment and habitat loss." The Island of Java is the most densely populated (by humans) island on Earth.

Passenger Pigeon

In 1866 in Ontario, Canada, *a single flock* of this species numbered approximately 3.5 billion, and took fourteen hours to pass overhead. Only locusts have larger groups or flocks. Deforestation and massive, mechanized over-hunting finished them off. The last pigeon perished in a zoo in 1914. The pigeon's plight can only be described as utterly catastrophic.

Zanzibar Leopard (Nearly Extinct)

Rural human inhabitants of Zanzibar widely believe this tiger is kept by witches who send the tigers out to harm and harass villagers. In 1964, an island-wide campaign was initiated to exterminate the demonized leopard. By the mid-1990s, authorities considered the species extinct, but several living Zanzibar leopards remain in existence.

Other than cloning, what can we do to preserve animals that are at risk of extinction?

Below is a *partial* list of endangered species from the Humane Research Council:

- Mountain gorilla (Central Africa)
- Leatherback Sea Turtle (Adults weigh 550-1500 lbs. and are the largest of all living turtles; found in most oceans.)
- Northern White Rhinoceros
- California Condor (Wingspan of nearly 10 feet; weight about 26 pounds. About 400 total Condors are known to exist in the wild and in captivity.)
- Brown Spider Monkey (Habitat: Colombia and Venezuela. Critically endangered.)
- Red Wolf (Weighs about 50 pounds; critically endangered. Habitat: Southeastern USA.)
- Chinese alligators (Only a few hundred still exist in ponds in the wild [but 10,000 are in captivity]. Habitat: Eastern China. Critically endangered.)
- Snow Leopard (Habitat: Central Asian Mountains. Effective population size [those likely to reproduce] is probably fewer than 2,500.)
- Giant panda (Habitat: Central China. About 270 are now in captivity, and between 1,600 and 3,000 exist in the wild.)
- Blue Whale (At 170 metric tons, and 75 feet in length, thee Blue Whale the *largest animal ever known to have lived.* Its heart weighs 1,300 pounds. Lives in all Earth's oceans. Estimates range from 5,000 – 12,000 blue whales still in existence. Endangered.)
- Tasmanian Devil (Habitat: Island of Tasmania, Australia. About the size of a small dog. Disease [cancer] recently reduced the population as much as

eighty percent. 10,000 to 15,000 devils survive in the wild.)

- Bonobo (Native to Congo Basin. Weighs 5 to 130 pounds. *One captive ape has a comprehension of 3,000 words.* Endangered. Population today: 30,000 to 50,000.)

The Humane Research Council performed statistical studies to extrapolate the number of *unknown* species from existing data on *known* species. Their findings indicate that perhaps about 89% of the Earth's living species have yet to be detected. Most of our oceans have yet to be fully explored.

Extinction and creation of species is natural, evolutionary process.

Human activity accounts for many of the recent extinctions of plants and animals. There certainly are some *natural causes* that have gravely harmed various species. Dutch elm disease and cancerous facial tumors that kill Tasmanian devils come to mind.

But *humanity* alone wins the dubious prize for wholesale, catastrophic ruination of far too many of Earth's species.

Urbanization destroys natural habitats. Deforestation creates human living space, croplands and wood for human homebuilding and other construction.

Global warming theories have numerous critics, but hard numbers and mountains of data can hardly be summarily dismissed. Climate change appears to be reaching huge areas of the world. After studying the Caribbean Basin and regions of South Africa, National Geographic News concluded that current carbon dioxide levels "eventually" would double in the two areas. This process might lead to the extinction of 56,000 plant species and 3,700 animal species in those areas alone. Is this fear mongering or a serious threat?

Finally, humanity itself may likely be an endangered species.

What if God gives up on the human beings occupying His beautiful Earth? He has good reason to abandon us. We have been murdering, raping, torturing and warring. We have damaged ecological balance by messing with GM and cloning. The first murder— Cain's killing Able—would be enough for me if I were God.

We should consider how we look in God's eyes. The rest of the animal world appears relatively well behaved. Sometimes I think we are changing His world is as if we are trying to play God. We make changes for our own selfish purposes. Centuries ago we worshipped false gods and images. Today we are acting as if we have His power—and His permission. We are merely specks in His universe, hardly worth any attention, but are a constant annoyance. Will God just shoo us away like we wave off a fly?

The scenario I describe below might be unthinkable, but I believe it is as plausible as it is implausible.

God could eliminate us in one second and restore Earth's natural ecology. You would not know it was coming or feel any pain—so do not worry. He would place zoo animals and house pets in natural habitats.

This "one-second event" is hard to imagine, and takes a lot of deep thinking and imagining. Only you can do this, on your own, alone. My own first thought would be one of regret—but I would be totally extinct and therefore would not be thinking about anything. This is just another abstract thought that we cannot fully comprehend—like a hereafter, eternity, and that our bodies and our empirical world are actually composed of almost-empty atoms.

Does anything I have written so far give you more

insight as to whether we evolved or were created? Either way, it would be God's work.

I lean more towards evolution because there are three distinct human races that are older than the written words in any books.

The Insect World

Like other opposites we have in our world—good/evil, night/day, and black/white—there are also good and bad insects.

Wikipedia defines insect ecology as the scientific study of how insects, individually or as a community, interact with the surrounding environment. These little critters can have an astonishing impact on ecology and a lasting effect on future years. It was not rats that spread the black plague, but the parasitical fleas they carried. When you look at old ships, you can see discs on the rope dock lines. The discs prevent rodents from getting aboard.

The 14th-century Black Plague killed about two-thirds of the European population, and still affects our own lives today. This 700-hundred year Butterfly Effect of tiny fleas had to be enormous.

I recommend Albert Camus' *The Plague,* a novel published in 1947. It describes how medical workers find commonality in their labor when Oran, Algeria is swept by bubonic plague. Camus' depiction of the impact of an epidemic on a single city—Oran— offers insight into how the Black Death must have impacted most of 14th century Europe, a huge area.

When I was in England some years ago, a new construction site revealed a very old grave full of the remains of bodies. Also in the grave were small smoking pipes made of clay (often called plague pipes) left by diggers who believed they could ward off the disease by smoking.

The Mosquito

The deadliest insect on earth is easily the mosquito. Annually, the anopheles mosquito causes 300-plus million malaria cases and one to three million human deaths. The Greek translation of anopheles is "useless." Mosquitos also carry canine heartworm and other deadly diseases.

Is there anything good that can be attributed to mosquitoes? No, answers Janet Fang for nature.com (July 2010). Some think mosquitos should be totally eradicated. "That sentiment is widely shared," writes Fang. "Malaria infects some 247 million people worldwide each year, and kills nearly one million. Mosquitoes cause a huge further medical and financial burden by spreading yellow fever, dengue fever, Japanese encephalitis, Rift Valley fever, Chikungunya virus and West Nile virus."

These flying pests have been around for more than 100 million years.

The Africanized Honey Bee

Also known as "killer bees," these insects were created when humans took African bees to Brazil in 1957. The idea was to breed an improved bee capable of producing more honey. Some of the new hybrid bees accidentally escaped.

Some claim these "bees from Brazil" produce more honey, but the Africanized bees are difficult to manage because of their extreme defensiveness. Their sting is no more potent than other varieties of honeybee.

Wikipedia: "What makes Africanized honeybees more dangerous is that they are more easily provoked, quick to swarm, attack in greater numbers, and pursue their victims for greater distances." Deaths from these hybrids remain relatively rare, but make headlines as the bees migrate northward into the United States. At least three B-movies about "killer bees" stoke fear in the general public.

This hybrid bee seems to be another sad chapter in humanity's attempt to "improve" on Mother Nature.

Below are five significant insects that have a positive effect on ecology:

Bees. A bee sting hurts and can be deadly to certain people. But the bee's value to our ecology is critically important. Bees are a major pollinator in the ecosystems of flowering plants and agriculture. Pollination benefits can be measured in the billions of dollars. Bees also provide honey and are respected universally for their reputation as hard workers.

Lacewings. In the lacewing's larvae stage, it eats small insects like aphids as well as the aphid's eggs. Aphids are a major cause of crop damage. The larvae lacewing is sometimes called the "aphid lion."

Ladybugs. In your garden they are a natural pesticide—one ladybug can eat 50 aphids per day. In the larvae stage, a ladybug can eat its own weight in 24 hours.

Parasitic Wasps. While they prey on agricultural insect "pests," these wasps have virtually no harmful effect on crops. Wasps are increasingly sold commercially at insectaries, and are then released directly onto croplands. The wasp's larvae can develop inside or outside the insect host. Wasps that kill their hosts are called parasitoids.

Praying Mantis. This bug simply looks like it is praying (see photo). They are ambush predators, and sometimes eat their siblings while young. The female tends to eat the male after mating. When I was a teenager in New Jersey, I was told that it was unlawful to kill them. This imaginary law obviously did not apply to the female.

Ecology and Related Science of the Future

We are not heading into the future with a clean slate. Severe damage has already been done and detrimental processes are still in progress. And there are activities we are starting now but they have not yet been identified regarding their potential hazards.

New Scientist (newscientist.com) names 25 environmental threats of the future. Below are several of them.

William Sutherland, a Zoologist at the University of Cambridge in the UK, told New Scientist "that the future supply of bio-fuel is already becoming a political issue because a thorough environmental assessment has yet to be carried out."

Offshore wind and wave power might be solutions to our energy predicament, but Sutherland and colleagues warn that these approaches could also affect marine ecosystems.

"In Australia, researchers have developed a novel way of controlling the invasive Red Fox—a virus that infects and sterilizes it—although it has not been released into the wild population. 'What happens if the virus spreads outside its target range,' asks Sutherland. 'Could it sterilize other foxes? Could the virus combine with another and infect different species?'"

In my own research I have uncovered far too many projects in process that could be listed, and therefore controlled. We have become a population of guinea pigs that are unaware of what we are eating, drinking, breathing, and medicating ourselves with, including health diagnosis processes. Even the economic crash caused by the banks and housing collapse (2006-2008) should have been known beforehand. Numerous analysts now acknowledge that they saw it coming.

Our government funds agencies to research future effects in order to protect our environment. One such program is Science to Achieve Results (STAR). Their vision: "To support the EPA to protect human health and the environment, the agency must have a base of sound science."

I lost count of all the government agencies involved with the environment (EPA, STAR, CERCLA, et al.). To me it seems that science and our government are going down two different roads with two different maps and two different vehicles.

The vehicle of science has a head start and a faster car.

As a business professional for most of my life, I have a persistent and troubling thought: If the government had to meet the goals that business has to meet, the government would be in bankruptcy. Yet the government regulates businesses and often does a poor job. I suspect we would be better off if the government took the back seat in science's vehicle. I am an Independent regarding politics and membership in organizations.

In Upton Sinclair's *The Jungle* (1906) our nation read about inhumane working conditions in Chicago's slaughter-houses. Sinclair's book also described outrageous contamination of food in packing plants. His primary goal was to show us how bad workers were treated. After the book was published and widely read, the government passed the Pure Food and Drug Act. Sinclair was dismayed. He said he had aimed at the public's heart, but hit it in the stomach.

Once again the government arrived late to the party.

The amazing breakthroughs in genetics since DNA was discovered are a bit eerie, particularly cloning of human organs. It is more science than fiction that our species will be cloned in the future. How will the cloning of animals affect the future of ecology on Earth?

What constitutes the process of cloning from DNA? Wiki: A sheep named Dolly "was born on 5 July 1996 to three mothers (one provided the egg, another the DNA and a third carried the cloned embryo to term). She was created using the technique of somatic cell nuclear transfer, where the cell nucleus from an adult cell is transferred into an unfertilized oocyte (developing egg cell) that has had its nucleus removed. The hybrid cell is then stimulated to divide by an electric shock, and when it develops into a blastocyst it is implanted in a surrogate mother."

Many scientists around the world are working on cloning, but more than countries, including the United States, have a ban on human cloning. But will the bans endure? And what about the other countries with no such restriction? I would not be surprised if successful human cloning were announced sooner rather than later. It may have already occurred. The subject is too intriguing for scientists not to face up to the challenge.

Two sci-fi movies came pretty close to the real thing: *The Boys from Brazil* (1978) and *A.I., Artificial Intelligence* (2001). Both dealt with cloning and created "human life" specimens.

In *Boys* Nazi's take preserved DNA from Adolph Hitler and use ova from surrogate mothers to replicate Hitler. In different countries, 94 clones are born with black hair, blue eyes, and minds like the Fuhrer's.

The movie *A.I.* is set in the future and reveals the technological power to create a life-like child android (robot). Although the child is not a clone, it demonstrates the drastic measures people are willing to take to "have" a child. This desire alone may lead to cloning. Throughout history, human minds have changed quickly when the need for satisfying a deep desire or longing presents itself.

This opens another question: Does a person's "soul"

affect their personality? Is the effect good, bad, or both? If we assume that the soul has an effect, would a cloned human have the same personality as its donor(s)? But if each person has a unique soul, then the cloned being likely would have a unique personality, different from the donor's. If the donor had a good soul and the cloned person an evil one, would that mean one was bound for heaven and the other for hell? This "soul" idea is very confusing. Does our brain have any influence on our soul? Or vice-versa? Remember that there is no mention in the Christian Bible of the duality of body and soul.

This leads to larger questions. What is the purpose of heaven and hell? I can understand that we want a heaven to exist. When I was a child, I first heard about heaven from Jehovah's Witnesses. I liked the idea and never gave it much thought. Today, after much research and thinking, I am led back to those larger unanswered questions.

* * * *

Some years ago I heard that scientists were looking for a way to extract the DNA from dinosaur eggs. Perhaps it was a topic on one of the TV science programs. The idea never left my mind so I did some research. According to the Institute for Creative Research (icr.org), "In order to clone dinosaurs, we need perfectly preserved and complete dinosaur DNA (which we don't have) and a living mother dinosaur to provide the living egg cell (which we also don't have and aren't going to have)." Thus the film *Jurassic Park* (1993) that grossed nearly a billion dollars was more fiction than real science.

In 2012, scientists from the U.K. implied that flatulent dinosaurs might have helped their own extinction. The colossal volume of methane gas from powerful farts from countless plant-eating dinosaurs may have raised the Earth's temperature 18 degrees Fahrenheit. Is this hypothetical contribution to global warming in the Mesozoic Age more fiction than science?

Would you personally really like to know your future a week, a month, or a year from today? Imagine that you and your best friend agree to go one year into the future from today with the understanding that you will be in the future for just 10 minutes and then return to present. After the experience, your friend is jubilant. He says that he got married to the girl he loved, had a job promotion and bought a new car.

You, on the other hand, have no memory of the future visit. You draw a complete blank as if you didn't exist. Does it mean you were dead? In a coma? You saw no heaven or hell—just nothing. You can't change the future, so what do you do now? You could die in one week, or in a few months. Are you sorry you tried this experience? How would you like this fate hanging over you? Not I!

The above time travel scenario is pure sci-fi—a thinking exercise about the future. Let's now travel to a planet 10 million light-years away. It is not possible, but pretend it is. We know from Einstein's theory that as you approach the speed of light, your body's clock slows down and stops at that speed. Scientists theorize that when you return back to Earth, you will be younger than when you left. I take issue, as a layman, with this theory. I intuitively and rationally believe that as we slow down to normal earth-time and speed, the time and aging process reverses itself. Anytime I see the infinity symbol (∞) in a math equation, I know there is a degree of uncertainty.

Nothing in the universe is as fast as light, but scientists at CERN (the European Organization for Nuclear Research) have built the largest machine in the world to reach that speed. It is called the Large Hadron Collider (LHC) which is in a tunnel 17 miles in circumference near Geneva, Switzerland. To date, the cost is $10 billion and rising. Its purpose is to address some of the most fundamental questions about physics. The scientists have designed the collider to accelerate tiny proton beams or lead nuclei (hadrons) to speeds of 99.9999991 of light and have them collide to reveal the particles created by the collision.

The LHC project began in 1998. Their first successful colliding occurred in 2012, when the LHC created a new heavy particle thought to exist fractions of a second after the Big Bang. They are attempting to see what really happened at the moment of the Big Bang.

When the LHC has its first successful collision, could it be like a supersize atomic bomb explosion? This is one more ecological project among many where we are not sure what the end result will be. If splitting the atom caused such a great release of power, what will the splitting of sub-atomic matter do?

These experiments are reminiscent of the inventor of dynamite, a Swedish chemist named Alfred Nobel. His invention, like Upton Sinclair's book, had unintended consequences. Nobel invented dynamite and other explosives, but was a pacifist by nature who simply enjoyed inventing things. He naively thought his dynamite would end all wars. Other people, of course, had different uses for the product. Nobel was eventually called "the merchant of death."

Alarmed at this epithet, Nobel's last will left most of his vast fortune to establish the "Nobel Prize." Over the past several decades, the Nobel Prize has been criticized for political bias and Euro-centrism.

Even when we attempt to do well for humanity, it can

blow up in our faces.

There have been many changes caused by inventors and scientists who addressed a current need at the expense of unknown future consequences.

Three powers of the universe can impact the ecology of the future: natural ecology (e.g., a meteor hits the Earth), humanity, and God. Of the three, only humanity changes the laws of nature. Natural ecology operates and evolves within the universal laws. As I have previously said, God does not interfere or we would just be puppets. This last idea is extremely difficult to accept.

I ask that you *think* it through, and choose to accept or reject it. If God is omnipresent, He is inside you and part of your ecology. However, you do your own thinking and make all your own choices, and therefore are accountable for your actions. When you go to church perhaps you can do unselfish and beneficial work to help others in God's name. That warm feeling you get may be God inside you (or you inside of God). I experience a sense of unity as I write this sentence and then look out my large window at the beauty of it all. Putting all science aside, I feel the pot of gold of life is simply to live it.

What happens to the warm, God-connected part of us when we die? I believe we have everlasting life because that part of us stays with God. This may be the "soul" most religions speak about. This may be the hereafter, the Alpha and the Omega, everything from A to Z.

Socrates taught that we are born already knowing everything and all knowledge is simply a matter of remembering. I don't believe this is true, but it is an interesting idea and Socrates was certainly a great thinker. Of course his world was relatively austere compared to ours.

Second, if God actually decided to end it all, it would

happen in a nanosecond. He could change the laws of gravity, the speed of light, prevent electrons from circulating around the atom, remove our atmosphere, and much more. We are here today only because He has willed it. If you were God, what changes would you make? Keep in mind that *every change causes change.* Are you capable of comprehending the total effect?

Third, future natural disasters could have devastating effects on the Earth and the United States in particular.

Palm Springs, California is subject to earthquakes. When I lived there about a decade ago, there was a lot of talk about when the Big One was coming. My house was close to the San Andreas Fault, which has a long history of earthquakes. Seismologists were certain the Big One could happen any day. Their instruments were showing how pressure was building up on the fault line. Within California, a major quake—magnitude 7.5 or greater—is highly probable within the next 30 years.

This ominous data affected me when I put my home on the market. Soon enough, my Realtor presented me with a purchase proposal. The offer was lower than my asking price. I told the Realtor I would think it over.

At the time, it was a buyer's market. The offer was all cash and higher than my original cost. I also had to consider that I did not have earthquake insurance—relatively few California residents bought it because of the high premium. As I was mulling over the offer in the evening, an earthquake hit and made small waves in my swimming pool. That tremor was all I needed to get me off the fence and accept the cash offer. Let's call it "a message from above."

Scientists believe the United States has many natural disasters just waiting to happen. A few may surprise you:

We are overdue for a Pacific Northwest Mega Quake.

It may register a whopping 9.0 or higher on the Richter scale. The fault is called the Cascadia Subduction Zone— and it's truly a tectonic time bomb. The fault runs parallel to the Pacific coast from northern California to Vancouver Island. It could rip open at any time.

The Yellowstone National Park's Super Volcano is a sleeping giant. This is another of nature's bombs just waiting to blow. Beginning three million years ago, there have been three "super eruptions." A caldera is land that has collapsed due to volcanic eruption. When one of Yellowstone's magma bubbles explodes some day, the ash may be 2,500 times more than Washington's Mount Saint Helens eruption in 1980.

With countless future events totally out of our control, we should clear our minds of them and enjoy one day at a time. The future is always one big question mark. We are best off when remembering Reinhold Niebuhr's serenity prayer:

God, grant me the serenity to accept the things I cannot change, the courage to change the things I can, and the wisdom to know the difference.

CHAPTER 7

YOUR UNOPENED GIFT

What makes you really different from everyone else on the planet? Is it a single trait or talent? Are you truly unique and "special?"

I know you are! Life experiences and the science of genetics prove you are.

In The Preface I state, "our lives from birth are predetermined by race, gender, religion, customs, cultures, social status, country of origin, and other rigid situations, but not ourselves." This chapter may help you take another look within yourself and question whether these conditions at birth are really who you want to be.

Just because identical twins possess the same DNA (they come from a single egg) does not make them *exactly* the same. As they go through life, the brain of each twin each will program itself as it reacts to various environments. It is highly unlikely each twin would have identical experiences. Even the twins' genes may be slightly different. There are cases where one twin is gay and the other is straight. A current twins study "is very different from traditional twin studies," says lead scientist Dr. Sven Bockland at UCLA. "Instead of simply calculating the role of genetics, we use ... gay-straight identical twin pairs to actually identify genes that play a role in sexual orientation."

If "identical" twins are different, then it is highly likely you must be original and exceptional, too.

More than any other part of our body, our brain makes us one of a kind. According to Dr. Dennis E. Coates (*Building Personal Strength*), "The part of the brain that produces the most striking differences in personality is the cortex, the outermost layer of brain matter. This is the thinking part of the brain where perception, language, learning, planning, problem solving, and most high-level functions are processed. Both your cortex and the thinking programs you acquire over a lifetime influence the thought patterns and behavior patterns we call *personality*. Since our brains aren't exactly alike and we don't all learn the same things, we think differently, act differently, and have different personalities. In this regard, every human being is completely unique."

Open the gift of *thinking* now and use your cortex to an extent you have not previously reached. You are indeed a unique person, and possess the resources to enjoy and release the magnificent power of *thinking* to reach higher and higher levels.

We all are familiar with fabulous mental giants like Edison and Einstein. Do you have passion for a goal that you have abandoned because *you* think it's impossible for *you* to achieve it?

I am often awed by the fast, precise, and brilliant conversation between actors in dramatic movies and plays. While watching I sometimes ask myself how the characters can think and respond so rapidly and cleverly in conversational give and take.

The answer, of course, is that perfect conversation rarely occurs in real life. Movie talk is carefully and professionally created. It is thought out and prepared by scriptwriters, then rehearsed and expertly delivered by skilled actors. Our everyday lives are likely never as perfect as the "let's pretend" movies. Movies re-present reality, and if the movie is good, we forget that what we are watching is merely an illusion. Yet in real life, we can have better conversations by practicing forethought and being better prepared.

A personal example of using careful thinking was on my first visit to Mainland China in 1984. I was a CEO of a global corporation in dire need of finding a lower cost source to manufacture our products. After doing my research, I settled on China, where our products could also be sold. I also identified the communist government's contacts in Beijing. I sent a fax to them in hopes of setting up an initial meeting to discuss my company's plans. I received an immediate response requesting that I come as soon as possible. We scheduled the meeting to occur in five weeks.

They were obviously eager if not downright hungry for our business. I quickly put in motion my rule of getting the facts and being fully prepared. I had five weeks to read books on their customs, history, and current events. I also read a few Chinese novels and spoke with recent Chinese immigrants to gain a sense of their traditional customs.

At our first meeting, I learned that money was their main interest, but because they were a noncapitalist nation, they could not relate to "profit." At this point I knew I would be doing most of the thinking for both sides. They did not know what their exact manufacturing cost was, and instead related only to the absolute money they would receive. They were simply following the number one rule of their government, which required very little thinking: *Get U.S. dollars.*

Over the next few years, there were more deals and more negotiations. My thinking evolved along with our growing business. New facts = new thinking.

Another tool I use in my life is to "edit" my thinking. When I approach a project or problem, I fully acknowledge that what I initially see and think personally may not be so important, especially over the long term. From training and experience, I know that what I do *not* see or do *not* think may ultimately matter the most.

I believe it is often useful *not* to believe most of what you read, see, or hear—it may be more theory than fact. More often than not, being a master skeptic will serve you well. I still have a vivid memory of being taught a "fact" in college that I later found was utterly false.

* * * *

The ultimate purpose of this book is to send you on your own thinking about the topics presented. I want to open your mind and make "you" ... *you*. How do you see yourself in the mirror of the vast universe?

The world as God-created is obvious, and the world you live within your mind is of your own making. Which one do you like the most? Or are you one of the few that have both worlds in synch and enjoy this tranquility of life?

In my research I found that there are over one hundred types of thinking, e.g., abstract thinking, logical thinking, visual thinking. To be amazed, simply Google "list of thought processes." There are myriad learned people out there engaging in fantastic mental gymnastics. However, many styles of thinking appear to be rather redundant, or just going through the motions. Playing patty-cake, so to speak.

Below, I offer several types of thinking and thinkers along with my comments. Try to see if some of them fit you.

• **Meditators and sleepers.** Meditators can quiet their minds for a period of time during the day. Complete relaxation, peace and quiet, and a refreshed mind allow creative juices to flow freely. Sleepers mentally describe a project or problem before they fall asleep, and let their subconscious mind come up with solutions in the morning.

• **Associative thinkers** compare and contrast, and try to find linkages and patterns among things that initially appear unrelated. How is *this* item like *that* item? Unknowingly, I did this all my life.

• **Visual thinkers** use art, doodles, photos, etc., to make visuals leading to The Big Picture, or the largest idea. Humans thought in images before they thought in words. Images have irresistible primordial power. Think *television*.

• **Storytellers** think in linear narratives, from a beginning to an end. Humans, after acquiring language, told myths or stories to explain the mysteries of existence.

• **Intuitive thinkers** rely on their "gut" feelings. They do not fear natural impulses and tendencies. They allow feelings to shape their creations from a rough idea.

• **Logical thinkers** are most comfortable with reason, deduction, and analysis. They can rearrange options in a project to show the true range and depth of choices available.

• **Prioritizers** are good at determining the relative importance of different facets of a project or problem. Think *triage*—who do we help or treat first when there are many people injured who all need some degree of care?

• **Organizers** are able to combine ideas into appropriate groups. A *notion* is a factoid or small bit of information. An *idea* is larger than a notion, and can explain a smaller set of notions. A good organizer is often "The Idea Person" in a group.

How many of the above categories fit you? I relate to at least several.

Actually I go further. I stretch out at least once a day on a favorite sofa and play with different ideas in a *proactive* way. I try to avoid un-fun *reactive* thinking. As a retiree, I have this luxury.

Crises happen in our lives when we have not thought about upcoming events or circumstances that could cause real pain and problems. Without preplanning for contingencies, we suddenly find ourselves *reacting* to something that has happened to us. We thrash and flail about trying frantically to think of a way to fix it. Sometimes we just give up and suffer. Proactive thinking—using foresight and vision—is the hallmark of beneficial and productive thinking.

In my business career, invariably there were certain workers who would think about all the reasons we should do a project. Another group of workers would think about all the reasons we should reject the project. They are in nearly every workplace: the positivists vs. negativists. Each group can offer necessary and useful input, but if the negative group consistently wins, the company will eventually decline or go out of business.

I am reasonably certain we can only be in one state of mind at a time. We can be happy or sad, depressed or cheerful, etc. If you choose to look at the positive things in life, you are prone to be a generally positive person. If you go around with a chip on your shoulder, you would tend to be often depressed. *This is the ecology of your brain and your attitude as you go through life.* A 180-degree conversion to a positive mood will improve your quality of life and decrease stress. The result is a healthier and happier you.

There are mental blocks to creative thinking as well as

healthy thinking skills. The brain needs exercising as much as the body to keep it in top shape and finely tuned.

Keep in mind that thinking does not necessarily always produce answers.

Happiness resides within your own mind. You can't wait for happiness to "just happen." You have to be proactive in personal ways that you know will open the door to satisfaction.

In a poem by James Russell Lowell, "The Vision of Sir Launfal," a knight dreams of fulfillment by trying to find the Holy Grail and fame and riches, but awakens to realize that true happiness is in giving and sharing right in his own home and neighborhood.

Various mental blocks are discussed in an excellent essay from by Brian Clark, founder of copyblogger.com.

Clark credits Roger von Oech's book, *A Whack on the Side of the Head* (2008), for inspiring him to write this blog post:

1. Trying to Find the "Right" Answer

One of the worst aspects of formal education is the focus on the correct answer to a particular question or problem. While this approach helps us function in society, it hurts creative thinking because real-life issues are ambiguous. There's often more than one "correct" answer, and the second one you come up with might be better than the first.

Many of the following mental blocks can be turned around to reveal ways to find more than one answer to any given problem. Try reframing the issue in several different ways in order to prompt different answers, and embrace answering inherently ambiguous questions in several different ways.

2. Logical Thinking

Not only is real life ambiguous, it's often illogical to the point of madness. While critical thinking skills based on logic are one of our main strengths in evaluating the feasibility of a creative idea, it's often the enemy of truly innovative thoughts in the first place.

One of the best ways to escape the constraints of your own logical mind is to think metaphorically. One of the reasons why metaphors work so well in communications is that we accept them as true without thinking about it. When you realize that "truth" is often symbolic, you'll often find that you are actually free to come up with alternatives.

3. Following Rules

One way to view creative thinking is to look at it as a destructive force. You're tearing away the often-arbitrary rules that others have set for you, and asking either "why" or "why not" whenever confronted with the way "everyone" does things.

This is easier said than done, since people will often defend the rules they follow even in the face of evidence that the rule doesn't work. People love to celebrate rebels like Richard Branson, but few seem brave enough to emulate him. Quit worshipping rule breakers and start breaking some rules.

4. Being Practical

Like logic, practicality is hugely important when it comes to execution, but often stifles innovative ideas before they can properly blossom. Don't allow the editor into the same room with your inner artist.

Try not to evaluate the actual feasibility of an approach until you've allowed it to exist on its own for a bit. Spend time asking "what if" as often as possible, and simply allow your imagination to go where it wants. You might just find yourself discovering a crazy idea that's so insanely practical that no one's thought of it before.

5. Play is Not Work

Allowing your mind to be at play is perhaps the most effective way to stimulate creative thinking, and yet many people disassociate play from work. These days, the people who can come up with great ideas and solutions are the most economically rewarded, while worker bees are often employed for the benefit of the creative thinkers.

You've heard the expression "work hard and play hard." All you have to realize is that they're the same thing to a creative thinker.

6. That's Not My Job

In an era of hyper-specialization, it's those who happily explore completely unrelated areas of life and knowledge who best see that everything is related. This goes back to what ad man Carl Ally said about creative persons—they want to be know-it-alls.

Sure, you've got to know the specialized stuff in your field, but if you view yourself as an explorer rather than a highly specialized cog in the machine, you'll run circles around the technical master in the success department.

7. Being a "Serious" Person

Most of what keeps us civilized boils down to conformity, consistency, shared values, and yes, thinking about things the same way everyone else does. There's nothing wrong with that necessarily, but if you can mentally accept that it's actually nothing more than groupthink that helps a society function, you can then give yourself permission to turn everything that's accepted upside down and shake out the illusions.

Leaders from Egyptian pharaohs to Chinese emperors and European royalty have consulted with fools, or court jesters, when faced with tough problems. The persona of the fool allowed the truth to be told, without the usual ramifications that might come with speaking blasphemy or challenging ingrained social conventions. Give yourself permission to be a fool and see things for what they really are.

8. Avoiding Ambiguity

We rationally realize that most every situation is ambiguous to some degree. And although dividing complex situations into black and white boxes can lead to disaster, we still do it. It's an innate characteristic of human psychology to desire certainty, but it's the creative thinker who rejects the false comfort of clarity when it's not really appropriate.

Ambiguity is your friend if you're looking to innovate. The fact that most people are uncomfortable exploring uncertainty gives you an advantage, as long as you can embrace ambiguity rather than run from it.

9. Being Wrong is Bad

We hate being wrong, and yet mistakes often teach us the most. Thomas Edison was wrong 1,800 times before getting the light bulb right. Edison's greatest strength was that he was not afraid to be wrong.

The best thing we do is to learn from our mistakes, but we have to free ourselves to make mistakes in the first place. Just try out your ideas and see what happens, take what you learn, and try something else. Ask yourself, what's the worst that can happen if I'm wrong? You'll often find the benefits of being wrong greatly outweigh the ramifications.

10. I'm Not Creative

Denying your own creativity is like denying you're a human being. We're all limitlessly creative, but only to the extent that we realize that we create our own limits with the way we think. If you tell yourself you're not creative, it becomes true. Stop that.

In that sense, awakening your own creativity is similar to the path reported by those who seek spiritual enlightenment. You're already enlightened, just like you're already creative, but you have to strip away all of your delusions before you can see it. Acknowledge that you're inherently creative, and then start tearing down the other barriers you've allowed to be created in your mind.

$$* \quad * \quad * \quad *$$

To further develop healthy thinking tools, consider Epictetus circa 60 AD: "[People] are not worried by things, but by their ideas about things. When we meet difficulties, we become anxious or troubled; let us not blame others, but rather ourselves ...our idea about things."

Also consider Proverbs 23:7: "As a man thinketh ... so is he."

There has been so much written over thousands of years about living a more comfortable life. It is all there for the taking, but sadly is seldom used.

Alfred Adler, founder of the School of Individual Psychology: "It is obvious that we are not influenced by 'facts' but by our interpretation of the facts."

As a young boy, I had a hard time learning to read and had to take remedial reading in my freshman class at high school. For me, words were too fuzzy and abstract, but math came easy. Who would have to argue the truth that 1 + 1 = 2? Obviously, I later conquered the reading problem. Borrowing from www.iaff.org on healthy thinking skills, below is a summary of some of the ways you can explore in order to change your negative moods to positive moods.

• Be precise in your thinking.

• Avoid words that are imperatives—always, never, should, must.

• React to what is real, not imagined.

• Consider the whole. Instead of focusing on a single negative detail about yourself or others, try to balance your view with the positive.

• Just because you feel something does not make it true. If you feel stupid, it does not follow that you are stupid.

• Fairness is relative, not absolute. Everyone has their

own judgment on fairness so don't expect all people to agree with you.

• It's not always about you. Looking for your self-worth in comparison to others is an empty exercise that will leave you thinking and feeling that you don't measure up.

You weren't born with these impediments to happiness and self-actualization; you picked them up as you went through life. You need a mental re-birth. Throughout this book I have stressed the need for you to be yourself and not something you were born into. It is not that you should stop believing what you were taught. But your decisions should be yours alone based on a solid knowledge of your range of choices.

All the ecosystems I have described try to maintain a natural and equal balance. Your mind also has to adapt naturally to feel comfortable with your fellow human beings and ecosystem. You will become happier, more accomplished, and have a zest for life. You will learn from your new beliefs how unique you are and gain the ambition to put them into action and make them work. You will increase your *output* of who you are in a world where most people are smothered with *input*. You may even find that you have a rebellious attitude towards those who have a typical herd-type mentality, but you can take pride that you are an individual.

Shakespeare, for all time, said it best in *Hamlet*: "To thine own self be true."

Go ahead and unwrap the gift—your mind's capacity has no limits. Bring it back to healthy thinking and discover the countless rewards in your future.

AFTERWORD

In the several million years of humanity's existence, any combination of tiny changes could have completely erased our presence—as an entire species, or perhaps just you as an individual. If the Black Death had claimed one of your unlucky ancestors, you would not be here now reading this sentence.

Despite the extraordinary odds against our surviving, some choose to be unappreciative or ignorant and hence make their lives seem meaningless. But others focus correctly on the enormous effect we can have on future generations. You and I are as much a part of ecology as ecology is of us.

If God had made some "changes" along the way, we might not be here. If God has the ability to see into the future, then He would already knows what effects His changes would make. Conversely, if He cannot see into the future, and makes changes that eventually cause harm, then it is He who made the future. I don't think He wants it that way.

The rational side of my mind says that He does not have both powers—the power to see into the future, and also the power to make changes right now—today. If God could see into the future and the Bible's Book of Revelation and Armageddon are true, it would create a very big question in my mind.

If God knows the world will end in a great conflict, would He interfere with our lives during our existence? Could He?

This is not a book on religion, but one that I hope you will use as you develop your own unique gift of *thinking*. I don't think God wants us not to use our rational minds to select how we see Him.

The American poet Adrienne Rich once wrote, "Responsibility to yourself means refusing to let others do your thinking, talking, and naming for you; it means learning to respect and use your own brains and instincts; hence, grappling with hard work." I absolutely believe this and cannot say it better.

I am in awe as I travel the world. The majestic beauty of the sky, geography, flora and fauna have always surprised and delighted me. Studying science and ecology helps make our planet seem even more astonishing. It reminds me of receiving an attractively wrapped present, only to be more pleased when I see what is inside. This is all the evidence I need to know God. No extra declarations, analyses, not even my own—are required.

For me, *The Ecology of Oneness* has certain similarities to the Christian Bible and perhaps other scriptural books. I notice that the Bible has many authors (some unknown), while my book has numerous references. Metaphors are used in both to help explain what is tough to understand. Many events in both books defy reality as we think we know it. Both books assert that God exists and is our Creator.

We know extremely little about the universe.

A prodigious puzzle is waiting to be solved. The answer(s) may likely be discovered when we ultimately learn to use the unopened parts of our brain. We have been given the gifts of survival and of potentially unlimited mental power. These two gifts are uniquely yours and should be fully treasured on your personal search. Take a fresh inventory of your life and its impact on the ecosystem in and around you.

In his 2005 book *The World Is Flat*, Thomas Friedman describes how barriers among nations continue to open up. International trade will hopefully promote increased harmony around the globe.

Isaiah 2:3-5: "[People] will beat their swords into plowshares and their spears into pruning hooks. Nation will not take up sword against nation, nor will they train for war anymore.... Let us walk in the light...."

Leaders of all modern religious faiths should be monotheistic. There is just one God. Mohammed united Arabia under Islam by overwhelming polytheistic tribes and bringing them into a monotheistic system of faith. In many cases, different belief systems had been a cause of continuous wars among tribes. Today's religions could do the same, and erase much of the disturbing animosity that currently exists.

I believe the majority of Earth's humans want world peace. The benefits would be enormous. Let us as whole— the human race—please God. He must already be pleased with many of us. Let us build on that and prove to Him how grateful we are for all the wonders He has bestowed upon us.